Published in 2010 by the
Addiction Technology Transfer Center (ATTC) National Office
University of Missouri - Kansas City
ATTC National Office
5100 Rockhill Rd.
Kansas City, MO 64110

At the time of publication, Pamela Hyde, J.D., served as the SAMHSA Administrator. H. Westley Clark, MD, JD, MPH, served as CSAT Director, Anne M. Herron, MA, served as Director of CSAT's Division of Services Improvement, Catherine D. Nugent, LCPC, served as the Quality Improvement and Branch Chief, and Donna Doolin, LSCSW, served as the CSAT Project Officer.

The first edition of *The Change Book: A Blueprint for Technology Transfer* was originally published in 2000. The second edition was originally published in 2002. The document published in 2010 is an exact reproduction of the 2002 edition content.

ATTC Unifying science, education
and services to transform lives.

AuthorHouse™
1663 Liberty Drive
Bloomington, IN 47403
www.authorhouse.com
Phone: 1-800-839-8640

First published by AuthorHouse 6/09/2010

ISBN: 978-1-4520-2736-4 (sc)

Library of Congress Control Number: 2010907867

Printed in the United States of America
Bloomington, Indiana

This book is printed on acid-free paper.

authorHOUSE®

TABLE OF CONTENTS

PREFACE 5

ACKNOWLEDGMENTS 7

CHAPTER ONE
What is Technology Transfer: The Principles 11

CHAPTER TWO
Creating A Blueprint for Change: The Ten Steps 15

CHAPTER THREE
Getting Started: Steps 1, 2 and 3 19

CHAPTER FOUR
Determining Your Targets: Steps 4 and 5 23

CHAPTER FIVE
Strategies and Activities: Steps 6 and 7 33

CHAPTER SIX
Implementation and Evaluation: Steps 8, 9 and 10 43

EPILOGUE by Barry Brown, PhD 49

ENDNOTES 51

OTHER RESOURCES 52

TOOLS FOR CHANGE 55

APPENDIX 57

ABOUT THE ATTC NETWORK 64

**THE CHANGE BOOK WORKBOOK
SECTION II, PAGES 1-25**

Web Resources

The 2nd edition of The Change Book has been updated and includes a list of needs assessments and readiness to change instruments, an annotated bibliography of seminal works in the field of technology transfer, research articles and links to pertinent Web sites. Please use this area to enhance your change initiatives, and to send us your ideas and stories about how you use The Change Book.

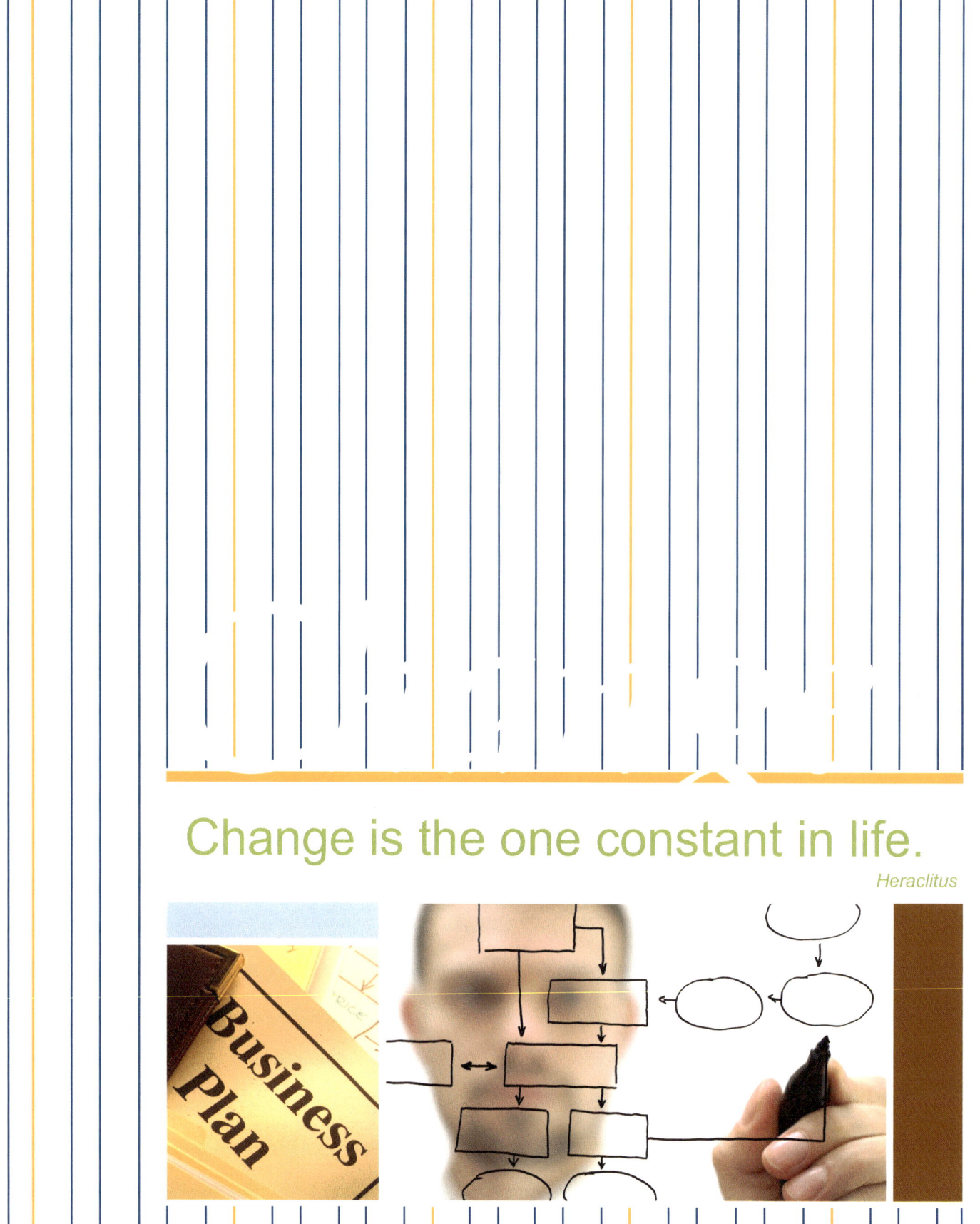

Change is the one constant in life.

Heraclitus

PREFACE

Although occasionally we like to try the new and different, on the whole, we humans resist change. We find comfort and a sense of confidence in the tried-and-true, in doing things the way we've always done them. Resistance to change is not unique to the individual. The groups, institutions and disciplines that we are part of also resist change. They often create barriers, sometimes inadvertently, for those within their ranks willing to embrace change. Change is often seen as a threat to stability.

Survival, however, dictates that in order to continue as a species, as a discipline, or as a profession, we must improve, adapt and constantly make the best use of what others learn and discover. Incorporating the new, however, requires changing how we do things. Research is constantly contributing new knowledge to the fields of substance abuse prevention and treatment. Yet this new knowledge often is not used in service settings.

Implementing changes based on research can be difficult for treatment agencies and professionals. Nonetheless, it is crucial to our field's health and that of our clients' that innovations on "how to best get the job done" become standard practice. That is what *technology transfer* is all about.

The Change Book: A Blueprint for Technology Transfer (The Change Book) was developed by the Addiction Technology Transfer Center (ATTC) National Office and the former ATTC Practice Committee. The word *technology* here is not limited to computers and electronics. Technology is used in the broader, more traditional sense of the word.

TECHNOLOGY:
the science of the application of knowledge to practical purposes; the application of scientific knowledge to practical purposes in a particular field.[1]

The Change Book is a tool to help you implement change initiatives that will improve prevention and treatment outcomes. It is designed for administrators, staff, educators and policy makers. Using this manual will increase your knowledge about effective technology transfer methods and will build your skills in implementing change within agencies.

The Change Book includes *Principles, Steps, Strategies* and *Activities* for achieving effective change. The *Principles* are required elements necessary for successful technology transfer, and the *Steps* are a guide to effectively creating and implementing a change plan. While the *Strategies* provide specific guidelines for working with individuals, groups and multiple levels within your organization, the *Activities* are actual tasks that can be used throughout the process to affect knowledge, skills and attitudes – ultimately creating behavior change. We encourage you to read this document in its entirety.

HISTORY OF
The Change Book

The Change Book was originally published in 2000. It was created as a follow-up to a November 1999 ATTC Technology Transfer Symposium. The event was designed to keep staff informed about what current research indicates works and doesn't work in technology transfer. Dialogue at the Symposium was rich, and some of the *Strategies* and tools presented in this guide were developed as a result of this discussion. Others evolved through a synthesis of recent research. Presenters at the Symposium included Jon Gold, Dennis McCarty, PhD, Thomas Valente, PhD, and Mary Marden Velasquez, PhD. Documentation of the Symposium proceedings is included as an appendix.

Since it was first published, *The Change Book* has proven to be a landmark document for the fields of substance abuse treatment and prevention. It was the first publication of its kind to outline the multidimensional aspects of instituting change specifically for addiction-related agencies.

Nearly 24,000 copies of *The Change Book* have been distributed since its first printing, and it has been downloaded nearly 16,000 times from the ATTC Network Web site. In addition, the document was translated into Spanish in 2002. Demand for this publication continues to outnumber supply, as requests for *The Change Book* are made daily.

WHO IS USING
The Change Book?

We have heard from a variety of people who have used *The Change Book* to implement change initiatives during the last four years. From frontline treatment practitioners using it to implement new treatment modalities in their agencies, to government officials using it within state departments to work toward system-wide changes, *The Change Book* is effectively guiding professionals across the country to create sustained change. We even heard from a neighborhood group who used *The Change Book* to solve a local trash problem.

ENHANCEMENTS
TO THE SECOND EDITION

Based on feedback from a number of users, we have made a variety of enhancements to the second edition of *The Change Book.*

Within *The Change Book* is a new list of assessment and readiness to change tools. An updated and more extensive list of related resources is also included at the end. We hope these enhancements and tools lead to even more successful change initiatives.

ACKNOWLEDGEMENTS

CONTRIBUTORS
TO THE FIRST EDITION

Publication of *The Change Book* required significant contributions from a number of people who deserve recognition. The vision was the product of a creative, energetic and spirited group of professionals that constituted the ATTC Practice Committee and the ATTC National Office (National Office).

At the time of printing the first edition, the former ATTC Practice Committee was one of seven national committees designed to serve the ATTC Network. Members worked to develop products and processes that would contribute to changes in professional practice and the larger treatment system. The Committee was comprised of representatives from several Centers within the Network, the National Office and selected experts representing multiple health and behavioral science disciplines and practice settings throughout the country.

We especially acknowledge Steve Gallon, PhD, chair of the ATTC Practice Committee and director of the Northwest Frontier ATTC for his vision and leadership. He skillfully set the stage for a dynamic, inclusive process resulting in the ATTC Technology Transfer Symposium and *The Change Book*.

ATTC PRACTICE COMMITTEE

At the time of the first edition, the following people comprised the ATTC Practice Committee and generously contributed to the development of this document.

Steve Gallon, PhD, Committee Chair
Northwest Frontier ATTC

Lonnetta Albright, Great Lakes ATTC

Jody Biscoe, MS, Texas ATTC

Roberto Delgado, MEd
Puerto Rico & U.S. Virgin Islands ATTC

Carlton Erickson, PhD, University of Texas

Patricia Fazzone, DNSc, University of Kansas

Gerald R. Garrett, PhD,
University of Massachusetts Boston

Sue Giles, MS, Mid-America ATTC

Jennifer Tate Giles, MSW, ATTC National Office

Laurent Javois,
Northwest Missouri Psychiatric Rehabilitation Center

Mary Beth Johnson, MSW, ATTC National Office

Richard Landis, MSW, DC/Delaware ATTC

Betty Singletary, LCDP, ATTC of New England

Pete Singleton, Mountain West ATTC

FROM THE ATTC
TECHNOLOGY TRANSFER SYMPOSIUM

One of the ATTC Practice Committee's primary projects was the planning and implementation of a November 1999 ATTC Technology Transfer Symposium which provided the foundation for the development of this document. The event was designed to keep staff across the country informed about what current research indicates works and doesn't work in technology transfer. Committee members collaborated on the design of the Symposium, planned and staged the event, and served as the editorial board for the project. The content of *The Change Book* came largely from the panel of experts who presented so admirably at this Symposium. Presenters included:

JON GOLD, branch chief of the Synthesis Branch of the Office of Evaluation, Scientific Analysis and Synthesis at the Center for Substance Abuse Treatment (CSAT). Gold presented an overview of CSAT's technology transfer initiatives and the role CSAT plays in the adoption of best practices within treatment systems.

(At the time of the second edition, Jon Gold is retired from CSAT.)

DENNIS MCCARTY, PHD, research professor at the Heller Graduate School for Advanced Studies in Social Welfare at Brandeis University. McCarty discussed the application of technology transfer strategies to individuals, organizations and systems targeted for change.

(At the time of the second edition, Dennis McCarty is a professor in the Department of Public Health and Preventive Medicine at Oregon Health Sciences University in Portland, Oregon.)

MARY MARDEN VELASQUEZ, PHD, associate professor in the Department of Family Practice and Community Medicine at the University of Texas-Houston Medical School. Velasquez discussed the Transtheoretical Model of Behavior Change, stages of change and the complexity of technology transfer at various stages.

(At the time of the second edition, Mary Marden Velasquez continues to teach at the University of Texas-Houston Medical School.)

THOMAS VALENTE, PHD, associate professor at the Population and Family Health Sciences Department in the School of Public Health at Johns Hopkins University. Valente summarized current research on diffusion of innovations and explained how to use opinion leaders in technology transfer.

(At the time of the second edition, Thomas Valente is director of the Master of Public Health program and an associate professor in the Department of Preventive Medicine in the Keck School of Medicine at the University of Southern California.)

JON GOLD, DENNIS MCCARTY, PHD, MARY MARDEN VELASQUEZ, PHD, and **THOMAS VALENTE, PHD,** all were delightful to work with, delivered eye-opening presentations and provided helpful editorial assistance as *The Change Book* began to take shape.

Participants at the ATTC Technology Transfer Symposium also contributed significantly to this publication. During the day-long event, they met to brainstorm issues and activities essential to the success of technology transfer initiatives. Many of their ideas have been included here.

OTHER NOTEWORTHY CONTRIBUTORS

First drafts of the document, the most difficult task in this kind of project, were developed by KELLY REINHARDT with support from the ATTC National Office. Kelly's contribution went beyond her excellent writing. She attended several ATTC Practice Committee meetings and shared insightful ideas while remaining open to a sometimes disparate variety of committee suggestions. Kelly's diplomacy, good nature, responsiveness, clear writing and timely submission of manuscripts deserve high praise.

EDNA TALBOY, MA, was responsible for creating an inspired organizational scheme and rewriting later drafts of the work. Edna's understanding of how to make a document maximally useful to the reader is evident.

The ATTC National Office kept communication flowing during the developmental process, wrote key chapters, provided other editorial input, and created the graphic and organizational design for the final product. ANGIE OLSON, MS, JENNIFER TATE GILES, MSW, and MARY BETH JOHNSON, MSW, the ATTC National Office Director, all played integral roles in developing this document.

BARRY BROWN, PHD, a leading researcher in technology transfer, was especially helpful in reviewing the document and writing the epilogue. His willingness to share his expertise, his enthusiasm for promoting technology transfer and his significant contribution to the final product are all greatly appreciated.

Special thanks to VALENTE, MCCARTY, VELASQUEZ and RON JACKSON who reviewed the first edition. Their support guided us through the final stages.

CONTRIBUTORS TO THE SECOND EDITION

A number of people have contributed to the enhancements and changes found in the second edition of *The Change Book*.

ATTC National Office staff and associates oversaw the development, organization and design of the second edition. The ATTC National Office Director, MARY BETH JOHNSON, MSW, CARLA INGRAM, CSACII, LCSW, JENNIFER ELLINGWOOD, MS, JENNIFER TATE GILES, MSW and ANGIE OLSON, MS all worked to create a user-friendly, cohesive edition that will benefit readers.

ATTC SERVICE IMPROVEMENT COMMITTEE

In addition, members of the ATTC Service Improvement Committee spent time reviewing and collecting resources and tools to enhance the second edition of *The Change Book*. They also reviewed the new edition and provided key input. Members of the committee included:

Michael Shafer, PhD, Committee Chair
Pacific Southwest ATTC

Carla Ingram, CSACII, LCSW, ATTC National Office

Cynthia Moreno Tuohy, NCACII, CCDCIII,
Central East ATTC

Michele Murphy-Smith, PhD, RN, Gulf Coast ATTC

Nancy Roget, MS, Mountain West ATTC

Pam Woll, MA, Great Lakes ATTC

Vonshurri Wrighten, MDiv, CACII, CCS,
Southeast ATTC

Jan Wrolstad, MDiv, Mid-America ATTC

You cannot step twice into the same river; for other waters are continually flowing in.

Plato

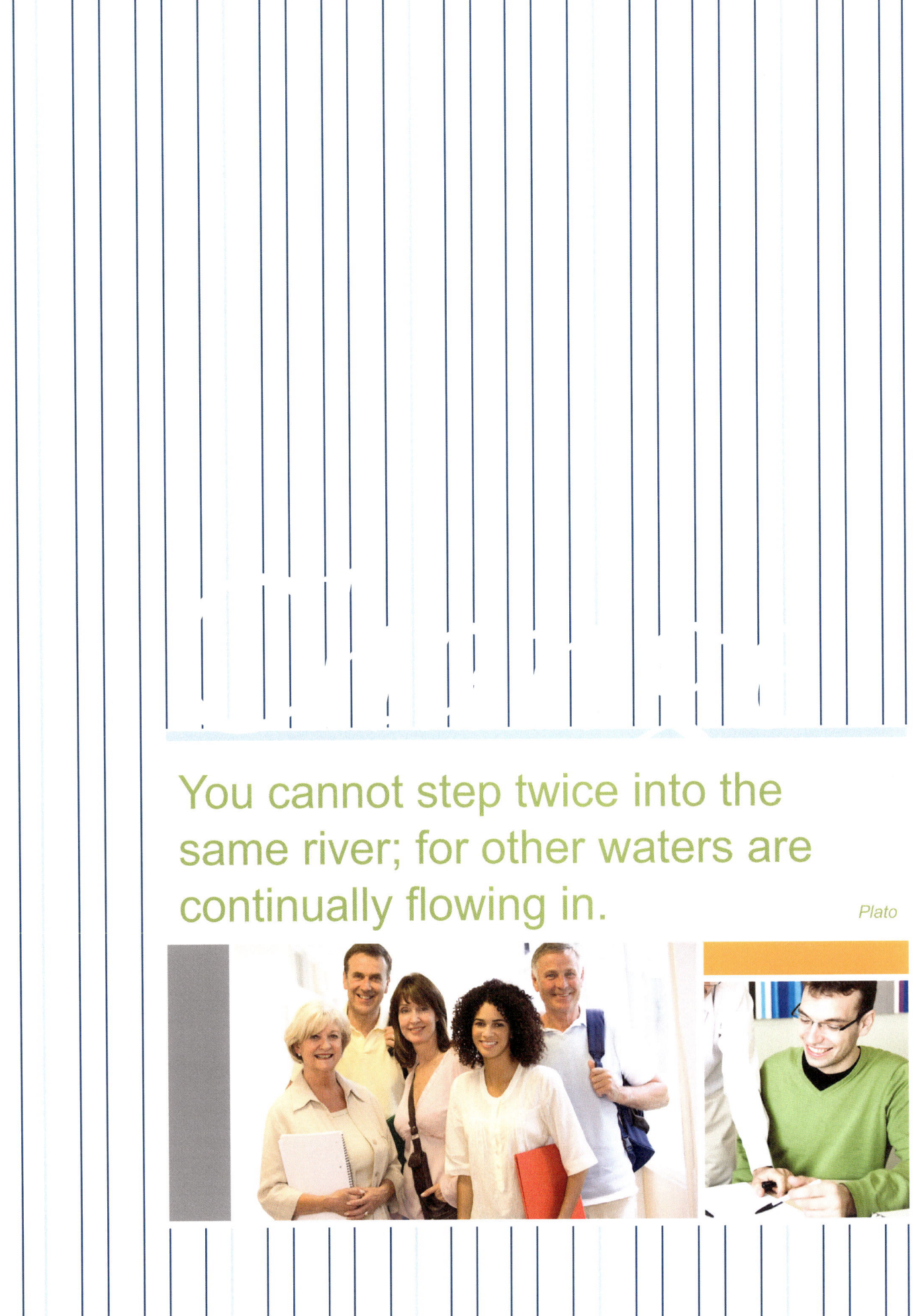

What is Technology Transfer and Why Is It So Important?

Technology by definition deals with the application of "scientific knowledge" to practical purposes in a particular field. In other words, technology deals with how we use the "tools of our trade" to do our job. In the treatment field, these tools fall into one of three broad classes: *knowledge, skills* and *attitudes*. The job of research is to constantly examine and evaluate these tools and any innovations or additions that occur over time.

TECHNOLOGY: the science of the application of knowledge to practical purposes; the application of scientific knowledge to practical purposes in a particular field.[1]

TRANSFER: to cause to pass from one person to another.[3]

Since technology changes over time, we depend on research to continually examine and evaluate technology changes for us. The technology used by our field provides answers to questions such as "how can prevention and treatment efforts yield better outcomes for clients?"

Given the mounting pressures to contain health care costs and the increasing emphasis on "outcome funding," entities connected to the prevention and treatment of substance use disorders have had to focus on improvements in practice that positively impact client outcomes. Yet there is mounting evidence indicating that much of the scientific knowledge gained from addiction-related research is *often not utilized in practice.* "There are more than 8,000 community-based treatment providers in the United States – and they account for the bulk of alcohol and other drug treatment. In spite of great strides made in research on the science and treatment of addiction, there are still many barriers to linking research findings with policy development and treatment implementation."[4]

So the question becomes, how do we transform what is useful into what is actually used? How do we move technology developed academically into standard professional practice? The answer is *technology transfer.*

Technology transfer is not new. Humans have been using technology transfer throughout our existence. In many ways, successful technology transfer is what determined which groups survived and which did not. This still holds true for disciplines and professions today.

TECHNOLOGY TRANSFER VERSUS TRAINING

Technology transfer is not simply passing on "how to best get the job done" to others in our field. That is training. Although training is one strategy in the technology transfer "tool box," too often brief flurries of training alone are thought to be sufficient in bringing about lasting change. The results are usually short-lived alterations in practice followed by discouragement and a return to familiar but less effective ways of doing things.

Technology transfer's scope is much broader than just training. It involves creating a mechanism by which a desired change is accepted, incorporated and reinforced at all levels of an organization or system. As Barry Brown, PhD, a leading researcher points out, "to produce behavior change, technology transfer strategies must not only develop the cognitive skills needed to implement a new treatment component, but may also have to induce or increase motivation for behavior change, reduce concerns about change generally, and/ or about the innovation specifically, and explore organizational issues in adopting new strategies."[5]

CREATING RESPONSIVE SYSTEMS

When beginning any change initiative, it is important to understand the multitude of factors that influence an agency's or an individual's willingness and readiness to change. Before we explore the specific *Steps* required to bring about and maintain change, we will examine more generally the factors influencing the success of any technology transfer initiative.

Effective technology transfer efforts require change at a variety of levels within the overall alcohol and drug treatment system – including clients, practitioners and agencies.[6] There will be barriers to change at each level and different *Strategies* required if practices within each level are to change. The challenge, according to Dennis McCarty, PhD, is finding *Strategies* to promote the adoption of new technology at the client/patient level, the practitioner/clinical level and the program/ organizational level. Targets for behavior change can also include the research community and policy makers.

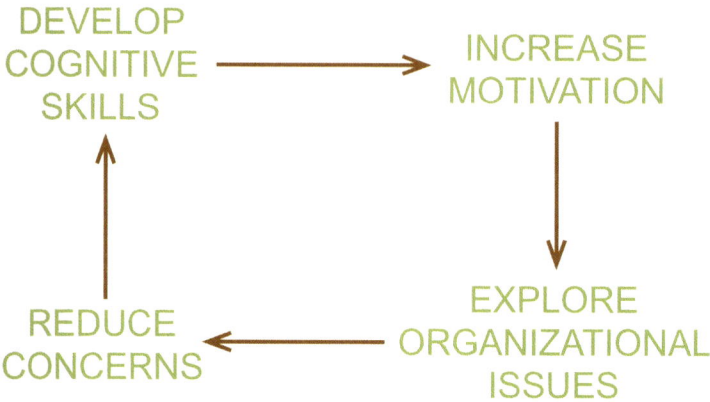

As Thomas Valente, PhD, reminds us, the introduction of practice standards or new innovations can have both positive and negative dimensions. On the plus side, they can improve practice, facilitate interchange and dialogue, standardize protocols and facilitate analysis. However, they can also upset existing procedures, destabilize structures and threaten the status quo.[7]

It is essential that system leaders be prepared to manage an expected level of resistance or tension as change elements are put into place. Specifically, it is important to anticipate attempts at all levels to maintain the status quo. Some individuals may feel threatened or uncomfortable in making changes that destabilize structures and upset existing procedures. If the strategy is sustained and based on proven evidence-based practices, resistance usually dissipates as familiarity with new procedures increases and improved client outcomes are realized.

In addition, the Transtheoretical Model of Change,[8] as presented by Mary Marden Velasquez, PhD, recognizes that organizations and individuals are often in various stages of readiness to change when presented with a technology transfer initiative. She stresses the importance of "marketing" an innovation through *Strategies* that correspond to each target's readiness for change.

Thus, the transfer of any technology in the prevention and treatment of substance use disorders requires the development of a broad range of competencies at multiple levels within the service delivery system. Creating a truly responsive system is not just a matter of developing proficiencies within direct service staff.

ADOPTION OF CHANGE
REQUIRES

> Policies that provide incentives for adopting innovative changes.
> System administrators knowledgeable and supportive of the proposed innovations.
> Agency directors willing to adapt their service designs to a new model.
> Supervisors skilled in implementing new practices.
> Opinion leaders who endorse the proposed system change. (See page 25 to learn more about opinion leaders.)
> Service providers with the knowledge, skills and attitudes consistent with the delivery of new practices.
> Opportunities for staff input and feedback.
> Opportunities for client input and feedback.

PRINCIPLES
OF EFFECTIVE
TECHNOLOGY TRANSFER

Studies of technology transfer in other disciplines and settings have identified several key *Principles* associated with success in the adoption of change. We can learn from these disciplines, as well as our own experience, which tells us that the following *Principles* must be incorporated into the change process for successful adoption to occur. A successful change initiative will be relevant, timely, clear, credible, multifaceted, continuous and will include active, bi-directional communication.

THE PRINCIPLES:
DEFINITIONS AND CHECKLIST

Below we have defined each of the *Principles* needed for successful technology transfer. When you begin designing a change initiative, ask yourself, "Is my plan relevant, timely, clear, credible, multifaceted, continuous and bi-directional?"

> RELEVANT

The technology in question must have obvious, practical application. *"The greater the relevance of research findings/technology to be transferred to the mission and goals of an organization, the more likely it is that those findings or technologies will be employed or adopted."* [9]

> TIMELY

Recipients must acknowledge the need for this technology now or in the very near future. *"Technology transfer is enhanced by the timeliness of the findings or technology in question to the decision-making process being undertaken by the organization."* [10]

> CLEAR

The language and process used to transfer the technology must be easily understood by the target audience. *"The language and format in which new technology or research findings are transmitted are critical to effective transfer."* [11]

> CREDIBLE

The target audience must have confidence in the proponents/sources of the technology. *"Use of research methodology is found to be central to the credibility of the technologies or findings to be transferred."* [12]

> MULTIFACETED

Technology transfer will require a variety of *Activities* and formats suited to the various targets of change. *Research on change initiatives indicates that the most effective Strategies are active rather than passive.* [13]

> CONTINUOUS

The new behavior must be continually reinforced at all levels until it becomes standard and then is maintained as such. *A sustained, comprehensive initiative that incorporates a variety of media and experiences is needed to support the desired change throughout the system over time.*

> BI-DIRECTIONAL

From the beginning of the change initiative, individuals targeted for change must be given opportunities to communicate directly with plan implementers. *Strong participation and active, bi-directional communication in a change initiative decreases resistance and increases buy-in to the change process.*

The chapters that follow will lead you in developing your own change initiative by exploring the *Steps, Activities* and *Strategies* required. Chapter 2 will explore the *Steps* generally and will review a hypothetical case study. Chapter 3 will focus on the foundational *Steps* 1, 2 and 3. Chapter 4 will focus on the assessment *Steps* 4 and 5. Chapter 5 will guide you through plan development *Steps* 6 and 7. Chapter 6 will explore plan implementation and evaluation in *Steps* 8, 9 and 10.

Creating a Blueprint for Change

The following ten *Steps* are your guide to creating a comprehensive blueprint for change. They will lead you through each aspect of the design, development, implementation, evaluation and revision of your plan. The next few chapters will show you how to create a change initiative that is suited specifically to your situation.

TEN STEPS
TO EFFECTIVE TECHNOLOGY TRANSFER

To change your agency or system from what it is now into what you want it to be, you'll need a blueprint to guide you. The Steps that follow provide a starting place. Some of the Steps may be worked simultaneously, or the order of the Steps may be changed to fit your needs. **For your plan to succeed, however, Steps 1-7 should be completed before you implement your change plan (Step 8).**

STEP 1
Identify the problem.

STEP 2
Organize a team for addressing the problem.

STEP 3
Identify the desired outcome.

STEP 4
Assess the organization or agency.

STEP 5
Assess the specific audience(s) to be targeted.

STEP 6
Identify the approach most likely to achieve the desired outcome.

STEP 7
Design action and maintenance plans for your change initiative.

STEP 8
Implement the action and maintenance plans for your change initiative.

STEP 9
Evaluate the progress of your change initiative.

STEP 10
Revise your action and maintenance plans based on evaluation results.

THERE IS A PAYOFF
FOR WORKING THE STEPS

Although effective technology transfer does not require technical expertise, it isn't as easy as we'd like to pretend. Working one's way through the *Steps* takes time, energy, patience and ingenuity. In the pages that follow, you will discover that in order to complete each *Step* you must first take smaller actions that require thought, investigation, negotiation, risk-taking, problem-solving and movement. But there is a payoff! By following the *Steps* to the letter, you will encounter less resistance, stronger participation, more voluntary adopters and less overall disruption to your program.

A CASE STUDY

To illustrate the *Steps* further, we will now examine a hypothetical technology transfer initiative. The following case study is a brief description of a change plan in action. Perhaps it will strike a familiar chord.

While reviewing a treatment outcomes report, the director of a community-based mental health and substance abuse center notices that many of the clients leaving treatment are diagnosed with more than one disorder. Not only are they dropping out at a higher rate than clients with only one diagnosis, but they are failing to "comply" with medical advice and then re-entering crisis center services at a rate three times higher than clients diagnosed with only one disorder. Concluding that the mental health and addiction treatment staff need to be educated/trained on co-occurring disorders, the director contacts a well-known professional training organization to assist in implementing a service modification within the agency.

The training organization representatives request that clinical supervisors participate in the assignment of frontline staff to training sessions stating that supervisor buy-in and support is critical. The director agrees and arranges to meet with supervisors and the training organization. At the scheduled meeting, the director rushes in twenty minutes late, introduces the trainers to supervisory staff and says she must leave due to an administrative emergency.

When the trainers explain that the training format will include a total of six two-hour sessions, the clinical supervisors express concern about how they are to "meet their service quotas" with staff gone so much. The trainers take these concerns back to the director, who assures them the matter will be taken care of and that course development should continue.

On the day before the first training session, the trainers request specific information about the participants. At this time they are told the participants have not yet been selected. The director pledges that frontline staff will be there the next day and insists that the training proceed as scheduled.

On the first day of training, only half of the anticipated participants file into the training room. Most are confused about why they've been mandated to participate, and voice annoyance at having been given less than one day's notice. All are concerned about how they will make adjustments to their work assignments.

WHAT DO YOU THINK?

Use the space below to reflect on the case study.

1. What was the problem? *(See Chapter 3 for information about questions 1-3.)*

   _____

   _____

2. How was a team organized to address the problem?

   _____

   _____

3. What was the desired outcome?

   _____

   _____

4. How was the organization assessed? *(See Chapter 4 for information about questions 4-5.)*

   _____

   _____

5. Who were the target audiences identified for change?

   _____

   _____

6. What approach was identified to be most likely to achieve the desired outcome?
   (See Chapter 5 for information about questions 6-7.)

   _____

   _____

7. What plans were made for implementing and maintaining the change initiative?

   _____

   _____

8. How were these plans implemented? *(See Chapter 6 for information about questions 8-10.)*

   _____

   _____

9. How was the change plan evaluated?

   _____

   _____

10. How was the change plan revised?

   _____

   _____

WHICH EFFECTIVE
TECHNOLOGY TRANSFER PRINCIPLES WERE SATISFIED?

Check which Principles *you think were satisfied in the previous case study.*

> RELEVANT
The technology in question must have obvious, practical application.

> TIMELY
Recipients must acknowledge the need for this technology now or in the very near future.

> CLEAR
The language and process used to transfer the technology must be easily understood by the target audience.

> CREDIBLE
The target audience must have confidence in the proponents/sources of the technology.

> MULTIFACETED
Technology transfer will require a variety of Activities and formats suited to the various targets of change.

> CONTINUOUS
The new behavior must be continually reinforced at all levels until it becomes standard and then is maintained as such.

> BIDIRECTIONAL
From the beginning of the change initiative, individuals targeted for change must be given opportunities to communicate directly with plan implementers.

In this chapter, we outlined the ten *Steps* to effective technology transfer and reviewed a hypothetical case study. In the coming chapters we will closely examine each of the *Steps* and will explore the questions that need to be answered to shape an effective change initiative.

Getting Started: Steps 1, 2 and 3

Below each of the ten *Steps*, you will find a series of questions that have been developed to help define this change initiative. These questions were designed based on the case study discussed in Chapter 2. These questions may be appropriate for your own change plan, or you may need to add to, delete or adapt the questions under each *Step* based on your own needs. To increase the likelihood of developing a successful change initiative, however, answer all the questions you include as completely as possible. Now we will closely examine *Steps* 1, 2 and 3, and will explore the questions that need to be answered to fully shape this change initiative.

WORKING THE STEPS

STEP 1   Identify the problem.

1. What is the issue or problem?
2. What data or other information support the existence of this issue or problem?
3. What is the current practice in your organization (for practitioners, administrators) that might be contributing to or maintaining this problem?

STEP 2   Organize a team for addressing the problem.

We encourage you to use a team approach from beginning to end with any change initiative. It is important to build your team with people from all levels of your agency. Your team's size will depend on the size of your organization and the particular change initiative you are implementing.

1. Who is affected by the problem (practitioners, administrators, clients, family members)? Do these individuals come from multiple disciplines (social workers, treatment counselors, mental health professionals)?
2. What do each of these groups think about the problem? Is there any perceived need to change by each of these groups? What do they think about each other?
3. Who are the opinion leaders within each of these groups? *(See page 25 for additional information on opinion leaders.)*
4. Who will your team members be?
5. How will you invite team members to participate in the change initiative?
6. When and where will you meet?
7. How will team members communicate (meetings, memos, listservs)?

8. How will you encourage and reward participation by team members (refreshments at meetings, recognition for participation)?

9. Are there people from outside your agency who should be involved in the change initiative (referral agencies, funders)?

STEP 3 Identify the desired outcome.

Be sure when defining your desired outcome to set goals and expectations at realistic and attainable levels. If your goals are too high and are not met, staff may become resistant to participating in future change projects.

1. What does current research show to be a realistic outcome for the problem? (Conduct a literature review in journals, on the Web, with government sources, etc.)

2. How have colleagues in similar organizations addressed the problem? What approaches have they used? What has been most effective? What outcomes have they achieved?

3. What do staff members think would be a realistic outcome for the problem?

4. Reflecting on this information, what will be your desired outcome?

APPLYING THE STEPS

To illustrate the application of *Steps* 1, 2 and 3 to a practical situation, we will use the case study presented in Chapter 2. You may have identified some of the following problems in the case study: only the director was aware of the problem, there wasn't a team formally identified to address the problem, and no specific outcome was identified. Now let's see how this change initiative might have unfolded if the implementers had followed the *Steps*.

STEP 1 Identify the problem.

1. **What is the issue or problem?**
   The administrator of a community mental health and substance abuse treatment center realizes that persons with co-occurring mental health and substance abuse disorders are having poor treatment outcomes.

2. **What data or other information support the existence of this issue or problem?**
   A treatment outcomes report indicates that a large number of clients with more than one disorder drop out of treatment, fail to "comply" with medical advice and re-enter crisis center services at a rate of three times that of clients with only one disorder.

3. **What is the current practice in your organization (for practitioners, administrators) that might be contributing to or maintaining this problem?**
   The agency is not actively employing any methods to address this problem. Clients are assigned to a case manager based solely on the presenting problem (e.g., substance abuse or mental health problem). Mental health counselors and substance abuse treatment counselors rarely have opportunities for interaction, and do not have any formal guidelines for working with dually diagnosed clients.

Organize a team for addressing the problem.

1. **Who is affected by the problem (practitioners, administrators, clients, family members)? Do these individuals come from multiple disciplines (social workers, treatment counselors, mental health professionals)?**
   The administrator of the agency, clinical supervisors, frontline mental health counselors, substance abuse treatment counselors, and clients and their families are all being impacted by this problem.

2. **What do each of these groups think about the problem? Is there any perceived need to change by each of these groups? What do they think about each other?**
   After reviewing the treatment outcomes report, most staff agree that something needs to change, but are unsure what they can do specifically. The mental health counselors and the substance abuse treatment counselors rarely interact or consult on cases, but because of the current situation some are beginning to meet informally. Family members have been aware of the problem, but have been unsure how to respond.

3. **Who are the opinion leaders within each of these groups?**
   The administrator used a combination of self-selection and staff-selection methods to identify opinion leaders among the clinical supervisors, frontline counselors, clients and family members.
   (See page 61 for information about opinion leader selection methods.)

4. **Who will your team members be?**
   The team will include the administrator of the agency, two clinical supervisors, two frontline mental health counselors, two substance abuse treatment counselors and a client with one year of sobriety and medical compliance. Also the chair of the Program Coordination Committee for the Center's Board will be kept informed and used as a consultant.

5. **How will you invite team members to participate in the change initiative?**
   The agency administrator will begin by circulating the treatment outcomes report to all staff. She will then invite the identified opinion leaders and other individuals from each level of the organization to participate. They will be asked to attend a special breakfast to discuss the issue.

6. **When and where will you meet?**
   At the initial breakfast, the team will be asked to meet every other week in the agency's conference room to work on developing and implementing a change initiative.

7. **How will team members communicate (meetings, memos, listservs)?**
   The team will communicate through meetings, memos and by telephone.

8. **How will you encourage and reward participation by team members (refreshments at meetings, recognition for participation)?**
   Refreshments will be provided at each team meeting, and team members will be recognized throughout the agency as "special task force" members. A celebration will be organized once the change plan is implemented.

9. **Are there people from outside your agency who should be involved in the change initiative (referral agencies, funders)?**
   No additional outside people will participate in this change initiative.

STEP 3   Identify the desired outcome.

1. **What does current research show to be a realistic outcome for the problem? (Conduct a literature review in journals, on the Web, with government sources, etc.)**
   Following the initial meeting, three team members were asked to research the problem. They have been given two hours a week (for three weeks) away from the office to conduct Web searches and literature reviews.

2. **How have colleagues in similar organizations addressed the problem? What approaches have they used? What has been most effective? What outcomes have they achieved?**
   Three team members were asked to call colleagues at similar agencies to discuss how they handle clients with co-occurring disorders. They asked the individuals what their outcome rates are with this population and what type of approaches they believe are most effective.

3. **What do staff members think would be a realistic outcome for the problem?**
   Because most staff have not been formally trained in treating dually diagnosed clients, many are unsure what a realistic outcome for the problem would be.

4. **Reflecting on this information, what will be your desired outcome?**
   Current research shows that agencies with a multidisciplinary team approach for treating dually diagnosed clients can expect to achieve on average 80% client compliance within the first year. Based on these findings and the advice of peers, staff have decided to set their desired outcome at 50% client compliance for year one and will work to reduce client re-entry to crisis center services by 25%. They wanted to set realistic, attainable goals for their initial change initiative. They agreed to reassess their target outcomes after year one.

> We have identified the problem, a team to address the problem and the desired outcome. Now we are ready for the next *Steps*. In Chapter 4, we will assess the organization and the specific target audience(s) for change.

Determining Your Targets: Steps 4 and 5

This chapter provides you with the mechanism for assessing your organization and the target audiences involved in your change initiative. *Steps* 4 and 5 give you information that is critical to subsequent planning, development and implementation. The information derived by completing *Steps* 4 and 5 will also help you establish whether your plan complies with the effective technology transfer *Principles* listed in Chapter 1. Remember, the *Principles* require that your plan be relevant, timely, clear, credible, multifaceted, continuous and include active, bi-directional communication.

ADDRESSING MULTIPLE LEVELS OF YOUR ORGANIZATION

(See Appendix, page 58.)

Which levels of the system will you target for change – clients, practitioners, the entire agency? *Strategies* and *Activities* chosen for your implementation plan will depend on whether you target the:

> Program/organizational level
> Practitioner/clinical level
> Client/patient level

Increasing the effectiveness of technology transfer efforts will require change at a variety of levels within the overall alcohol and other drug treatment system. If you do not address *consumer and client concerns*, your change initiative will be working at cross-purposes with the very people you hope will benefit most. Without addressing the needs of both the *supervisors and clinical staff*, changes will not be incorporated into practice. At best you will get perfunctory compliance and eventually staff will return to doing things the way they were done in the past. Without targeting the *agency's management*, you will not have the key funding, incentive and structural support needed to implement and maintain change. Your initiative will exist only on the periphery of everyday life within the organization.

INCREASING THE EFFECTIVENESS OF TECHNOLOGY TRANSFER EFFORTS WILL REQUIRE CHANGE AT A VARIETY OF LEVELS WITHIN THE OVERALL . . . TREATMENT SYSTEM.

BARRIERS TO CHANGE

As in all professional fields, there are barriers to change at each organizational level. When assessing each level, it is important to realize that even if all the *Principles* of effective technology transfer are taken into account, barriers may still arise. They are real and cannot be ignored. Therefore, we must be prepared to implement well thought-out, realistic *Strategies* for addressing them. Remember, each barrier offers an opportunity for change.

In this section we will examine the barriers and opportunities within the system structure, policy makers, research community, agency treatment staff and client population.

REMEMBER TO OUTLINE YOUR OWN LOCAL BARRIERS AND OPPORTUNITIES IN THE CHANGE BOOK WORKBOOK.

SYSTEM STRUCTURE

THE BARRIER: *Federal, state and local government entities and individual agencies charged with responsibility for the prevention and treatment of substance use disorders are fragmented, don't communicate, and often work at cross-purposes.*

THE OPPORTUNITY: *These systems provide fertile ground for change efforts such as cross-training initiatives that improve client outcomes and increase cross-system collaborations.*

THE POLICY MAKERS

THE BARRIER: *Community-based treatment agencies often receive federal, state, health insurance and private funds. These funding sources may not support or may be in conflict about funding innovative research-based treatment methods. Public or payor policies may not support the application of new scientific discoveries, especially when they challenge established and familiar practices and beliefs.*

THE OPPORTUNITY: *Community organizations collaborating with researchers are ideally positioned to educate policy makers about the efficacy of research-based methodologies.*

THE RESEARCH COMMUNITY

THE BARRIER: *Most scientific research is rewarded by publication in professional journals. These journals are often not available to the clinical practice community because journal subscriptions can be costly and tend to be written for scientific audiences. Formal training for clinicians seldom includes practical lessons in using research literature to improve and change practice. Even when available, research reports typically do not meet the practical needs of frontline staff wanting to apply new research findings.*[11]

THE OPPORTUNITY: *Increase the occasions for dialogue between researchers and treatment organizations to heighten awareness about the need for and benefit of collaborative relationships between the two groups.*

AGENCY STAFF

THE BARRIER: *Staff can be resistant to change due to many factors including: a lack of understanding the new information, a lack of incentive for change, competing priorities, funding limitations, fear of failure and a general fear of change.*

THE OPPORTUNITY: *When properly addressed, opportunities may include cross-training of staff, open dialogue between clinicians and administrators about research, and a heightened awareness of improved client outcomes throughout an agency.*

THE CLIENT POPULATION

THE BARRIER: *People in drug and alcohol treatment tend to have a fairly high incidence of relapse, have high levels of co-existing disorders, and often face social problems such as unemployment or homelessness. This creates a population difficult to track and treat for extended periods of time.*

THE OPPORTUNITY: *Because our clients face such desperate situations, they are often willing to try new treatment options.*

In addition to these barriers and opportunities, there are a number of other factors that may influence the rate at which change is adopted within an agency including: the size of the agency (small versus large), the type of work setting (community-based treatment program, medical or mental health program, freestanding clinic), staff composition (number of staff in recovery, staff with certification/licensure, staff with master's degrees, culture and ethnicity of staff), learning styles, receptivity and commitment of staff to the change initiative, and the resources available to implement change. There is differing research about the impact these factors will have on the rate at which change occurs. We mention them here as food for thought when you are designing and assessing your change plan.

OPINION LEADERS
(See Appendix, page 60.)

In addition to being prepared to address the barriers affecting change initiatives, we must also remember that people adopt change at different rates – some early, some later. A person's beliefs about a change concept prior to a technology transfer effort will influence the rate at which they adopt the change.

Recent research indicates that *opinion leaders* are particularly influential in the adoption of change by others. Who are the opinion leaders in your target audiences? Although these individuals are selected by the group as "leaders," they are not necessarily the leaders shown on an organizational chart. They influence the group through their attitudes and behaviors. Opinion leaders tend to be conservative, uphold the "norms" of the group and often wait to see where a group is going before adopting something new.

Valente identifies nine methods that can be used for selecting opinion leaders. Of these nine, the four most easily implemented in agency settings include: self-selection, self-identification, staff-selection and the positional approach. *(See page 61 for more information about these methods.)*

It is important to determine the opinion leaders in each of your target audiences because the perception of approval by peers, active encouragement by peers and perceived support by opinion leaders for a new behavior are all factors that influence the adoption of change.

STAGES OF CHANGE
(See Appendix, page 62.)

In addition to targeting your initiative to the many levels of an organization and utilizing opinion leaders in the change process, there are five "stages of change" that can influence the outcome of technology transfer efforts. Each stage depicts a different degree of readiness for change within clients, staff and an agency as a whole. Because the effectiveness of a particular strategy depends on the target audience's degree of readiness, it is important to determine where on the continuum your target audience is. If you have multiple target audiences, they are very likely to be at different stages in the change process. In developing a change plan *(Steps 6-7)*, you can choose *Strategies* that have been shown to be effective at various stages of change. The five stages of change are:[15]

EACH STAGE DEPICTS A DIFFERENT DEGREE OF READINESS FOR CHANGE WITHIN CLIENTS, STAFF AND AN AGENCY AS A WHOLE.

PRECONTEMPLATION
People and organizations are not thinking about change. They think, "Everything is working like it is supposed to."

CONTEMPLATION
People and organizations are thinking about change, but often have ambivalent thoughts or feelings. They think, "It might be a good idea to change, but is the situation really that bad?"

PREPARATION
People and organizations are getting ready to make a change, but they are not yet ready to act. They think, "Something has to change if we are going to fix this problem."

ACTION
People and organizations are actively changing. They think, "We are changing our practice by _____."

MAINTENANCE
People and organizations have already made a change and are working to maintain the new behavior. They think, "How is this change working? How could we improve our change plan?"

VALUING AND ADDRESSING
RESISTANCE TO CHANGE

Change can be and often is very stressful. Resistance at all levels of the organization should be expected and will require attention. Not everyone will understand the value in making a change. Therefore, it is important to thoroughly explain what the payoff can be for them personally and for the organization as a whole. Let staff know how making a change can save them time, enhance their skills and benefit clients.

People do not intentionally resist change, but they often resist having change imposed upon them, especially when it is done hastily and without consideration. This is why it is important to include as many people as you can in the change process at the very beginning – from identifying and understanding the problem – to understanding and being actively involved in the solution.

In some situations, resistance can be valuable to your change efforts. Resistance often springs from legitimate needs and/or doubts. By "rolling" with resistance, you can identify legitimate issues within your change plan that truly need revisiting or more planning. Addressing resistance directly will lessen the likelihood that opposition will spread and influence others.

It is important to provide a forum where fears and concerns can be expressed freely and without consequences. Recognize that people experience change in different ways. *Education, communication and participation are the keys to reducing fears and building trust.* Emphasize the pros and cons of the change initiative and share decisional factors openly. Let those affected know that the change process is dynamic and will evolve based on their feedback.

Use your own experience as a human being who has undergone change to understand and anticipate the wariness and uncertainty with which most people face change. A wise change agent allows people to express their discomfort, hesitation and doubts. You may not be able to eliminate resistance, but you can understand it, respect its foundations, and remove some of the reasons for it.

> BY ROLLING WITH RESISTANCE, YOU CAN IDENTIFY LEGITIMATE ISSUES WITHIN YOUR CHANGE PLAN THAT TRULY NEED REVISITING OR MORE PLANNING.

TIPS FOR MINIMIZING
RESISTANCE

- > Directly address resistance
- > Discuss pros and cons openly
- > Provide incentives and rewards
- > Actively involve as many people as possible from the beginning

- > Listen to fears and concerns
- > Educate and communicate
- > Develop realistic goals
- > Emphasize that feedback will shape the change process

- > Actively listen to resistors
- > Celebrate small victories
- > Use opinion leaders and early adopters for training and promotion

Some groups and individuals cannot be won over, but their opposition and their resistance can be neutralized. With those individuals who seek negative attention, focusing on their resistance will only reward and encourage it. In these cases tolerance may be a more effective strategy. Once you reward the early adopters, the nay-sayers might envy those rewards and decide to join the change process.

If you offend stakeholders, it is important to make amends as fairly and honestly as you can, and keep the change process moving. You can often win over more than a few stakeholders with openness, authenticity, flexibility and responsiveness to their needs and concerns.

It takes time and energy to work through significant changes, whether in the workplace or in our personal lives. Many times resistance to change is a natural reaction of people trying to understand what is expected of them and how the change will impact their lives.

A WISE CHANGE AGENT ALLOWS PEOPLE TO EXPRESS THEIR DISCOMFORT, HESITATION AND DOUBTS.

WORKING THE STEPS

Now it is time to apply what we have learned about assessing organizations and target audiences. *Steps* 4 and 5 will guide us through this process. Remember to add to, delete or adapt the questions under each *Step* when creating your own change initiative.

STEP 4   Assess the organization or agency.

1. What is the existing organizational structure and size of your agency?
2. What is the mission of the organization?
3. What type of work setting is it (medical, substance abuse treatment, mental health, freestanding clinic)?
4. What is the staff composition (administrators, supervisors, counselors)?
5. What is the education and experience level of staff?
6. What is the cultural makeup of the staff and/or clients?
7. What are some of the organizational barriers to change (funding, physical structure, organizational structure, policies)?

8. What are the organizational supports for implementing change (strong desire for better outcomes, identified opinion leaders, available funding)?
9. At what stage of change is the organization operating with regard to this change initiative (precontemplation, contemplation, preparation, action, maintenance)?
10. Where will the resources come from to provide support for the change initiative (funding, community support, internal support from counselors and clients)?
11. What will the adoption of this change mean at all levels of the organization? What are the benefits for administrators, supervisors and counselors?
12. What things are already happening that might lay the foundation for the desired change?

STEP 5   Assess the specific audience(s) to be targeted.

1. Who will be targeted for the desired change (administrators, supervisors, counselors, clients)?
2. Are there any incentives to change (for counselors, supervisors or the entire organization)?
3. What are the barriers to change (for counselors, supervisors or the entire organization)?
4. At what stage of change are *each* of these target audiences (administrators, supervisors, counselors, clients)?
5. How will the practice(s) of those involved be affected by change?
6. Can we identify the opinion leaders within each of these target groups?
7. What additional support will the target audience(s) need to bring about change (e.g., training, policy changes, financial, additional personnel)?

APPLYING THE STEPS

In the case study from Chapter 2, organizational factors were not appropriately taken into account or assessed, and there was no identification of who should be targeted for change. Now let's see how this scenario would have unfolded if *Steps* 4 and 5 had been applied.

STEP 4   Assess the organization or agency.

1. What is the existing organizational structure and size of your agency?

The agency is a not-for-profit organization with an administrator responsible to a board of directors. Currently there is an agency hierarchy consisting of one administrator and 15 clinical staff.

2. **What is the mission of the organization?**

   The mission of the organization is to provide outpatient mental health and substance abuse treatment services for men, women and families.

3. **What type of work setting is it (medical, substance abuse treatment, mental health, freestanding clinic)?**

   The work setting is a freestanding outpatient community mental health and substance abuse treatment center.

4. **What is the staff composition (administrators, supervisors, counselors)?**

   The staff consists of one administrator, three clinical supervisors, three substance abuse counselors and nine mental health counselors.

5. **What is the education and experience level of staff?**

   The administrator and clinical supervisors all have master's degrees in related fields. One of the three substance abuse counselors has a master's degree, one is working on a bachelor's degree and one is in recovery with little formal training in substance abuse treatment. Five of the mental health counselors have bachelor's degrees, one has a master's degree, two are working on a master's degree and one has little formal training in counseling with a high school education.

6. **What is the cultural makeup of the staff and/or clients?**

   The majority of clients are male (71%), between the ages of 25 and 40. The ethnic makeup of these clients is African-American (61%) and Caucasian (39%). Eleven staff are Caucasian and four are African-American and range in age from 22 to 50. One-fourth of staff are male and the rest are female.

7. **What are some of the organizational barriers to change (funding, physical structure, organizational structure, policies)?**

   The organization receives its funding from the State Departments of Psychiatric Services and Substance Abuse. It also receives some funds through fundraising efforts by the board of directors. The physical structure of the building doesn't support strong communication among any staff. Substance abuse treatment counselors and mental health counselors are placed on opposite sides of the building. The clinical supervisors and administrator are in separate areas of the building from frontline staff.

8. **What are the organizational supports for implementing change (strong desire for better outcomes, identified opinion leaders, available funding)?**

   The administrator and the board president recently attended a conference about treating patients with dual disorders and shared their enthusiasm with clinical supervisors and the board about improving the agency's outcomes with this population. In addition, the administrator is familiar with the services her local ATTC provides and believes they can provide technical assistance to support change within her organization.

9. **At what stage of change is the organization operating with regard to this change initiative (precontemplation, contemplation, preparation, action, maintenance)?**

   As an organization, change is not something that has happened often. Many staff have been at the agency for a number of years and are comfortable with the systems in place. As new staff members are hired straight from undergraduate and graduate level programs, however, current research and concepts are introduced. Because the agency administrator tries to remain "current" in her knowledge of trends in the field, she tries to lead the agency by being open to new concepts. An organizational readiness for change instrument was administered by two graduate students from the public administration program at a local university. They determined that overall this agency is in the contemplation stage.

10. **Where will the resources come from to provide support for the change initiative (funding, community support, internal support from counselors and clients)?**

    The administrator will use 70% of the agency's annual training budget for the training, materials, reports and posters. The administrator is also requesting additional training funds from a local foundation, and the board has agreed to raise $5,000 for the training effort. In lieu of monetary compensation for overtime, staff will be given compensatory time off for overtime spent developing this initiative.

11. **What will the adoption of this change mean at all levels of the organization? What are the benefits for administrators, supervisors and counselors?**

    Changing the way staff currently work will affect everyone – administrators, supervisors, counselors and clients. The benefits to changing include: new skills for counselors, improved outcomes for clients, improved morale for staff and possibly stronger recognition of the agency within the community which could lead to increased funding.

12. **What things are already happening that might lay the foundation for the desired change?**

    The administration is aware of the problem and is gathering information to improve outcomes with clients.

STEP 5   Assess the specific audience(s) to be targeted.

1. **Who will be targeted for the desired change (administrators, supervisors, counselors, clients)?**

   Frontline counselors (both mental health and substance abuse) and clinical supervisors will be the two audiences targeted for change.

2. **Are there any incentives to change (for counselors, supervisors or the entire organization)?**

   Recognition of staff members participating in the change initiative, and temporarily reduced caseloads for those willing to pilot the new recommended approach are possible incentives to change.

3. **What are the barriers to change (for counselors, supervisors or the entire organization)?**

   A barrier to change for the counselors is that they have been unaware of the problem and are just beginning to see a need to change. The policies and procedures of the agency have been in place for a number of years and no one has reviewed the way this population is handled at the agency. Most staff are comfortable with their daily routines and are not aware of what current research says about working with persons who have dual disorders. Mental health counselors and substance abuse treatment counselors have different treatment paradigms.

4. **At what stage of change are *each* of these target audiences (administrators, supervisors, counselors, clients)?**

   Because the administrator and supervisors have been discussing the issue for awhile, they are ready for change to occur, but aren't clear about what changes should take place. They are in the preparation stage. Most counselors are now aware that there is an issue and are considering changes. They are in the contemplation stage of change.

5. **How will the practice(s) of those involved be affected by change?**

   The daily routine and work of counselors and supervisors will be affected by a change in policies.

6. **Can we identify the opinion leaders within each of these target groups?**

   The administrator identified opinion leaders while organizing the team (Step 2).

7. **What additional support will the target audience(s) need to bring about change (e.g., training, policy changes, financial, additional personnel)?**

   It is possible that training about treating persons with dual disorders, policy changes in current systems and technical assistance from the local ATTC will all be needed to bring about change.

At this point, we have identified the problem to be addressed and determined the desired outcome. We have also assessed the organization and specific audience(s) within the organization to target for change. Based on this critical analysis, in the next chapter we will use *Steps* 6 and 7 to identify the recommended approach needed, design an implementation process and develop a plan for maintaining change.

Thinking Through Your Plan of Action

Once your target audiences have been identified and stages of change for each have been assessed, *Steps* 6 and 7 will help you develop your "plan of action." *Step* 7 focuses specifically on the *Strategies* and *Activities* you'll use. These will be dictated by the information you obtained in *Steps* 4 and 5. Some *Strategies* are unique to a particular level. Others, such as providing evidence of what a change can accomplish, are common to all.

STRATEGIES FOR AFFECTING CHANGE
AT MULTIPLE LEVELS OF AN ORGANIZATION
(See Appendix, page 58.)

As we discussed in Chapter 4, it is important to target many levels of an organization when planning a change initiative. Below are strategies for affecting change at the program/organizational level, practitioner/clinical level and the client/patient level.

PROGRAM/ORGANIZATIONAL LEVEL
When addressing this level, it is important to:

1. Provide evidence of how the recommended approach works.
2. Inform agencies and organizations that although financing is important, it cannot and should not be the basis for deciding whether to address a needed change. Many changes can be made with limited time and finances.
3. Secure the tangible support (financial or other) of stakeholders and funders who have policy making authority such as a single state agency, grantor, board, etc.
4. Acknowledge and respond to the concerns or barriers perceived by the agency or organization.
5. Develop training and diffusion *Strategies* that are suited and will appeal to each of the target groups that make up the organization.

PRACTITIONER/CLINICAL LEVEL
When addressing this level, it is important to:

1. Provide evidence of how the recommended approach works.
2. Educate the practitioner about the approach.
3. Refer to the effectiveness of related or parallel technologies in other areas or fields.
4. Provide incentives for clinicians to use a recommended approach (peer support, financial incentives, outcomes monitoring).
5. Identify early adopters and allow them to model the new behavior.
6. Utilize a multifaceted approach.
7. Utilize advertising and marketing to get the word out to staff.

CLIENT/PATIENT LEVEL
When addressing this level, it is important to:

1. Provide evidence of how a recommended approach works.
2. Educate the client/patient about the approach.
3. Refer to the effectiveness of related or parallel technologies in other areas or fields.
4. Utilize advertising and marketing to get the word out to clients.

STRATEGIES TO USE
DURING EACH STAGE OF CHANGE
(See Appendix, page 62.)

Not only is it important to utilize *Strategies* appropriate for each level of an organization, but we must also take into account the extent to which our target level is ready for change. Suggested *Strategies* appropriate to use at each stage of change are listed below.

PRECONTEMPLATION
People and organizations in this stage are not thinking about change. They believe that everything is working fine.

1. Raise the awareness of this group about the approach under consideration.
2. Use a variety of media to disseminate information.
3. Make multiple attempts to disseminate information.
4. Conduct a needs assessment. Evaluate current practices and share results.
5. Recognize that people and organizations are at this stage of change for different reasons.
6. Assess the decisional balance and elicit conversation regarding the benefits versus the drawbacks about making a change.

CONTEMPLATION

People and organizations thinking about change can be overwhelmed with too much
information. They need just enough to stimulate their interest and curiosity.

1. Provide "tastes" of the topic to build interest.
2. Provide evidence for the effectiveness of a recommended approach – don't just provide statistics.
3. Probe the group to learn their reasons for concern.
4. Build self-efficacy: a person's belief in his or her ability to carry out or succeed with a specific task. Treatment professionals need this as much as clients do.
5. "Tip" the decisional balance. Help to identify more pros than cons about the recommended approach to build confidence in the initiative and move people toward change.

PREPARATION

Movement to the "action" stage of change is not always a smooth one. It is important to prepare individuals
for change.

1. Be sure the language and format of the information you disseminate are clear to your target audiences.
2. Have your target audience(s) assist in the development of the change plan.
3. Make sure the change can be adopted in your particular setting.
4. Remove any site-specific barriers to implementation.

ACTION

It is important to support people and organizations in change. We often focus on getting people and
organizations to buy into change, but then withdraw support once the action stage is reached. Leaders of the
change must provide resources to continue the change initiative over time.

1. Provide information in a user-friendly fashion.
2. Encourage questions and problem-solving.
3. Have frequent interpersonal contact. Mentoring during this stage is important.
4. Provide ongoing monitoring.
5. Offer nonthreatening feedback.

MAINTENANCE

It is important to support new behavior so people and organizations follow through and don't just move on to
the next "innovation."

1. Continue communication (updates, newsletters, Web sites, listservs, telephone trees).
2. Continue interpersonal contact.
3. Encourage communication and problem-solving.
4. Develop skills to maintain the behavior.

DECIDING WHAT TO DO

Choosing a *Strategy* implies choosing subsequent actions or *Activities* that can be used to transfer knowledge and information in order to bring about a change in behavior.

YOUR SELECTION OF SPECIFIC ACTIVITIES WILL BE AFFECTED BY:

1. Your thorough evaluation of the problem.
2. The audience(s) targeted for change.
3. The stages of change at which these target audiences are operating.
4. The availability of opinion leaders to influence change.
5. The resources available for your change initiative.

ACTIVITIES

IDEA LIST

The following "idea list" provides suggestions for *Activities*, that when used in combination, assist in the change process. Keep in mind that effective technology transfer is not one-dimensional, and therefore cannot include only one *Activity*. Some *Activities* can be implemented agency-wide, others will be used one-on-one with individuals. This list is not exhaustive, but rather should stimulate your thinking about which combination of *Activities* might work best in a given situation.

ADMINISTRATIVE/ STRUCTURAL ACTIVITIES
> Develop strategic plans
> Implement legal and funding mandates
> Implement policy changes
> Provide on-site technical assistance
> Provide rewards/incentives for change (intrinsic or extrinsic)

PERSON-TO-PERSON ACTIVITIES
> Conduct mentoring
> Encourage peer-to-peer coaching
> Provide clinical supervision
> Use early adopter influence
> Use opinion leader influence
> Utilize role playing

EVALUATION ACTIVITIES
> Collect baseline data
> Conduct needs assessments
> Conduct outcome/impact studies
> Conduct process evaluation
> Develop reports

EDUCATIONAL ACTIVITIES
> College courses
> Conference workshops
> Education groups within your agency
> Lectures
> Online courses
> Professional meetings
> Quizzes and examinations
> Self-directed learning packages
> Short training courses (1-5 days/topic specific)
> Workshop training sessions

INFORMATION DISSEMINATION ACTIVITIES
> Ads and public service announcements
> Audiotapes
> Books/manuals
> Curriculum packages
> E-zines (online magazines)
> Fact sheets
> Government publications
> Internal reports with results/ accomplishments
> Memos
> Newsletter articles
> Posters
> Press releases
> Professional journal articles
> Promotional flyers
> Teleconferences
> Video instruction
> Web sites

WORKING THE STEPS

Next we will use *Step* 6 to identify the most appropriate approach to achieve the desired outcome. Then in *Step* 7, we will determine exactly how to implement and maintain our change plan. Remember to add to, delete or adapt the questions under each *Step* when creating your own change initiative.

STEP 6 Identify the approach most likely to achieve the desired outcome.

1. What approach does research indicate to be effective in addressing the problem? (Again, conduct a literature review in journals, on the Web, with government sources, etc.)
2. How have colleagues in other organizations addressed similar problems? What has been most effective? What approaches have they used?
3. What do staff members think is an appropriate approach to reach the desired outcome?
4. Reflecting on the information obtained, what is the desired approach you've identified?
5. What are your reasons for selecting this particular recommended approach?

STEP 7 Design action and maintenance plans for your change initiative.

1. Based on the stages of change, what *Strategies* and *Activities* do you think will work best for each organizational level you plan to address?
2. What is the timeline for your change initiative?
3. What are the resources needed to implement these *Strategies* and *Activities* (e.g., funding for training, staff time, paper and printing)?
4. Who will be responsible for implementing the specific *Strategies* and *Activities*?
5. How will the logistics be handled (e.g., memos, gathering baseline data, scheduling training)?
6. How will you collect, analyze and report baseline data? Will you use an assessment? Do you have a computer? What resources are available for this process?
7. How will you include those affected by the change in the change process (invite counselors into planning sessions, solicit client opinions, invite input from community partners, board members, family members)?
8. What evidence will be presented to the target audience(s) to support the desired change?
9. How will the pros and cons of adopting the recommended approach – perceived and real – be presented (to clients, practitioners and administrators)?
10. What *Activities* will be employed to maintain the technology transfer initiative (quarterly progress meetings, monthly reports on progress toward outcomes)?
11. What resources are needed to implement and maintain this initiative?

APPLYING THE STEPS

In our case study from Chapter 2, the desired outcome was vague. The administrator didn't make informed decisions about the appropriate approach to use. The action plan only included one activity (training), and no maintenance plan was developed. Now let's continue applying the *Steps* to the case study for a revised scenario.

STEP 6 Identify the approach most likely to achieve the desired outcome.

1. **What approach does research indicate to be effective in addressing the problem? (Again conduct a literature review in journals, on the Web, with government sources, etc.)**

   The initial literature reviews conducted by team members identified a primary methodology that worked most effectively with this population – a multidisciplinary counseling approach. After consulting with each other, staff continued searching for additional research about this methodology.

2. **How have colleagues in other organizations addressed similar problems? What has been most effective? What approaches have they used?**

   During the survey process with colleagues, team members asked those who were having positive outcomes with this population what methods they were employing. It appeared that the organizations having more successful outcomes had professionals from multiple disciplines working together to address clients with co-occurring disorders.

3. **What do staff members think is an appropriate approach to reach the desired outcome?**

   Three team members reviewed client files and determined that the agency's clients in this population with the most successful outcomes had been assigned to both a mental health counselor and substance abuse treatment counselor. Treatment retention rates were significantly higher with this method.

   Two team members also conducted phone interviews with staff about what approaches might achieve better outcomes with this population. They identified a number of ideas including cross-training for staff, weekly "check-in" meetings to communicate about clients, changing the assessment process, and conducting groups for clients with co-occurring disorders.

   In addition, two team members surveyed a sample of clients with co-occurring disorders and their family members to get their thoughts on the problem. Feedback concluded that most don't think the counselors ever talk to each other or really understand all of the things the clients are going through.

4. **Reflecting on the information obtained, what is the desired approach you have identified?**

   Based on all the information collected, the team has determined that a multidisciplinary integrated treatment approach would be most appropriate for the agency to implement.

5. What are your reasons for selecting this particular recommended approach?

Several factors have indicated that this approach is appropriate and is likely to improve outcomes — the research, other colleagues having positive outcomes with similar clients, previous client records and feedback from staff and clients.

STEP 7 Design action and maintenance plans for your change initiative.

1. Based on the stages of change, what *Strategies* and *Activities* do you think will work best for each organizational level you plan to address?

Based on results of the stages of change assessment, the majority of counselors are in the contemplation stage of change. Because people in the contemplation stage respond well to "tastes" of topics and evidence favoring the change, the following Activities will be used. Fact sheets, research articles and memos will all be distributed on a regular basis about the recommended approach. Posters will be hung in key areas of the agency. Because opinion leaders have been included in the taskforce, their influence will be utilized to encourage change throughout the agency. Open discussions about the pros and cons of using a multidisciplinary team approach will also be utilized with counselors.

Supervisors and the administrator have been identified as being in the preparation stage of change. Thus, it is important to include them in the planning stages of the initiative to ensure their continued "buy-in."

Four half-day training sessions will be presented for counselors and supervisors about the benefits of using multidisciplinary teams to treat persons with co-occurring disorders. The trainings will be given on-site by a local training agency so staff are able to attend.

A pilot program will be conducted with members of the taskforce to begin employing new methods of addressing co-occurring clients. Two mental health counselors and a substance abuse counselor will work together to treat the same dually diagnosed clients. They will meet regularly to discuss clients, and will work together in assessing client progress. One clinical supervisor will oversee their work.

2. What is the timeline for your change initiative?

The change will be initiated throughout the agency over a period of nine months. During the first month, planning and research will be conducted by the taskforce.

In the second month, taskforce members will begin introducing change concepts about the recommended approach to the rest of the staff. In month three, the taskforce will distribute information about the pros and cons of the change concept and will hold the first on-site training.

In month four, the pilot team will begin employing new methods of addressing co-occurring clients for three months. Data will be collected based on this initial pilot. Also in month four, the second on-site training will be held for all staff. Clients in the pilot will also be interviewed to determine their thoughts about the changes taking place.

In month five, the pilot team will share results and experiences with other staff. Two additional trainings will be held. Information will continue to be distributed throughout the agency.

In month six, the change initiative will be evaluated. If results are positive, the multidisciplinary approach will be expanded to staff throughout the agency. Once a month, progress will be charted and shared with all staff. After the first three months that all staff use the new approach, an "evaluate our success" party will be held. Changes will be made to the change plan as needed based on evaluation and feedback.

3. **What are the resources needed to implement these *Strategies* and *Activities* (e.g., funding for training, staff time, paper and printing)?**

   Staff time will be needed to develop memos and reports, provide briefings and attend training sessions. Computer access will be needed to do research, and money will be needed to provide the four on-site trainings and refreshments for taskforce meetings.

4. **Who will be responsible for implementing the specific *Strategies* and *Activities*?**

   Taskforce members are responsible for distributing information throughout the agency and conducting planning and feedback sessions with staff. Supervisors are responsible for overseeing the pilot study and collecting baseline data. The administrator is responsible for authorizing funding and arranging the training sessions.

5. **How will the logistics be handled (e.g., memos, gathering baseline data, scheduling training)?**

   The taskforce members will develop a logistical plan with an accompanying timeline.

6. **How will you collect, analyze and report baseline data? Will you use an assessment? Do you have a computer? What resources are available for this process?**

   In this instance, an outcomes report was used to identify the problem. Outcome reports are currently produced by an outside consultant for the agency on an annual basis. The data is collected from reports prepared by clinical supervisors each quarter indicating client progress. Client charts from the pilot study and second implementation phase will be used to collect additional data. Taskforce members will take the lead in designing an instrument for collecting this data.

7. **How will you include those affected by the change in the change process (invite counselors into planning sessions, solicit client opinions, invite input from community partners, board members, family members)?**

   The counselors and supervisors affected by the change will be involved in the implementation process by answering surveys about the type of training and information they would like to receive. They will also be asked to bring whatever information they have about working effectively with co-occurring disordered clients to planning sessions with taskforce members to be used in designing the initiative. As the initiative progresses, memos and updates will keep counselors and supervisors informed. Clients and family members will be asked to provide feedback about how services could be enhanced within the agency.

8. **What evidence will be presented to the target audience(s) to support the desired change?**

The decisional factors (the research, experience of similar agencies, etc.) that went into developing the change plan will be distributed in an easy-to-read, brief memo to all staff at the beginning of the change initiative, and then again as it progresses. A poster displaying these decisional factors will be displayed in the conference room and lunch room.

9. **How will the pros and cons of adopting the recommended approach – perceived and real – be presented (to clients, practitioners and administrators)?**

Based on open discussions between staff, a list of pros and cons for the change initiative will be generated and sent to all staff. Taskforce members will also be encouraged to talk to their teams about the pros and cons of the initiative. These discussions will emphasize that the change process is dynamic and will reflect the experiences and feedback provided by staff, clients and family members.

Counselors will tell clients that a change is taking place in the agency so staff can be more receptive to their needs. Clients will be encouraged to provide feedback in writing or verbally to counselors about the changes taking place.

10. **What *Activities* will be employed to maintain the technology transfer initiative (quarterly progress meetings, monthly reports on progress toward outcomes)?**

Regular updates from the administrator about the pilot, monthly written reports from members of the taskforce and the "evaluate our success" party will all be used to keep staff informed and knowledgeable about the change initiative. Six, nine and twelve month trainings on treating clients with co-occurring disorders will also be scheduled.

11. **What resources are needed to implement and maintain this initiative?**

Staff time, training funds, needs assessment tools, and organizational and individual stages of change assessments will all be needed.

We have now identified the problem to be addressed and determined the desired outcome. We have also assessed the organization and specific audience(s) targeted for change. Based on this critical analysis, we then identified the recommended approach needed and designed an implementation process for change. Finally, we developed a plan for maintaining change within our organization. Now it is time to put these plans into action, evaluate how it's going, and make necessary revisions. Chapter 6 will focus on *Steps* 8, 9 and 10 which will guide us through this process.

From out of all the many particulars comes oneness,and out of oneness come all the many particulars. *Heraclitus*

Implementation and Evaluation:
Steps 8, 9 and 10

By the time you have reached this point in your blueprint, the foundation will be complete. The institutions and people that will create and benefit from this change, and the tools you've selected as most suitable for your plan will all be identified. Now, let's focus on the specifics needed to take your plan from an idea into reality.

KEEP YOUR EYE ON THE BALL . . .
THINGS RARELY GO AS PLANNED

Change is an especially dynamic process. In order to reach our goal we must be willing to take frequent, clear, hard looks at our progress. We must assess if changes or modifications need to be made to our plans and then revise and adapt based on this assessment. Unfortunately, this is easier to say than do.

We need to keep in mind that success is NOT the perfection of the plan as originally drawn, but the reaching of the goal or outcome we set in *Step* 3. We need to institute mechanisms that will force us to evaluate our efforts and progress so that we aren't wasting time, energy, resources and the goodwill of others. These mechanisms will assist us in revising, adapting and moving forward.

MOTIVATION

Because change probably will not happen quickly, it is important to celebrate small successes as they happen. Reward your "early adopters" for their participation and use them to promote the change initiative throughout your agency. Like opinion leaders, their support for the initiative is likely to influence others to change. It is also important to provide opportunities for feedback from all parties involved and incorporate this feedback into your plan of action.

SUCCESS IS NOT THE PERFECTION OF THE CHANGE PLAN AS ORIGINALLY DRAWN, BUT THE REACHING OF THE GOAL OR OUTCOME WE SET IN *STEP* 3.

WORKING THE STEPS

Now, it is time for *Step* 8, which is where we put *Steps* 1 through 7 into action. It is imperative to remember that while some of the previous *Steps* may be worked simultaneously, or the order of the *Steps* may be changed to fit your needs, **for your plan to succeed *Steps* 1-7 should be completed before you move on to *Step* 8.** *Steps* 9 and 10 will guide us through evaluating our change initiative and revising our plans based on results.

STEP 8   Implement the action and maintenance plans for your change initiative.

Now it is time for you to put Steps *1 through 7 into action!*

STEP 9   Evaluate the progress of your change initiative.

You'll use information collected in Step 9 *to determine if changes to your action and maintenance plans need to be made (Step 10).*

1. As you begin implementing the change initiative, what is the initial feedback from your target audience(s)? What is the reaction to print materials, training, online courses, etc.?
2. From the client, staff or administrative perspective, what adjustments need to be made to your plan?
3. Have the objectives of your change initiative been met? What is the impact of your efforts?
4. How will you share the results of your change initiative with frontline staff, supervisors, administrators, the research community?
5. How will you celebrate successes/results and support continuous feedback?

STEP 10   Revise your action and maintenance plans based on evaluation results.

Now it is time to revise your current change plans based on the information you collected in Step 9. *Once you have decided which revisions to make, you can continue the change process.*

1. How will you incorporate evaluation feedback into your plans?
2. How will you address resistance to the change initiative?

APPLYING THE STEPS

In the case study in Chapter 2, the logistics of the change plan were poorly implemented for the only activity planned (training), and there was no evaluation process in place. Thus, there was no way to revise the action plan based on evaluation results. Now let's complete our revised scenario by applying *Steps* 8, 9 and 10.

STEP 8   Implement the action and maintenance plans for your change initiative.

Now it is time for you to put Steps 1-7 into action!

STEP 9   Evaluate the progress of your change initiative.

Once you have developed a plan for your change initiative, it is important to evaluate throughout the process.

1. **As you begin implementing the change initiative, what is the initial feedback from your target audience(s)? What is the reaction to print materials, training, online courses, etc.?**

   Initial feedback about the change initiative is generally positive. Comments from counselors include, "I didn't realize just how many clients had not been returning to treatment," and "I appreciate the opportunity to share my expertise and concerns about this problem."

   Others, however, are more resistant. Their comments include, "I don't understand the problem. Why should I have to change my practice?" and "This only relates to other staff members. It doesn't really impact me."

   Overall, staff participating in the pilot are experiencing improved morale and are enjoying having the opportunity to communicate more often about clients. The substance abuse counselors and mental health counselors are pleased to have peer input about their clients, and find that they are working together more often even when it doesn't pertain to the pilot study.

2. **From the client, staff or administrative perspective, what adjustments need to be made to your plan?**

   Clients would like a way to voice their opinions confidentially in writing on a regular basis, and staff have requested that they have more opportunities to learn about current research. The late adopters would like an opportunity to discuss their position regarding the change initiative.

3. **Have the objectives of your change initiative been met? What is the impact of your efforts?**

   Evaluation results indicate that clients are staying in treatment longer. Results are minimal and slow, but they are positive and show a gradual progression toward the desired outcome/goal.

4. **How will you share the results of your change initiative with frontline staff, supervisors, administrators, the research community?**

   Once early adopters in the pilot program begin experiencing success with clients, they will share their successes with other staff at monthly staff meetings and in daily conversations. Once a month, a one-page, colorful graph and memo will be sent to every staff member in the agency depicting progress toward the goal. These memos will also include personal notes of encouragement written by the administrator.

5. How will you celebrate successes/results and support continuous feedback?

The "evaluate our success" party will be held to celebrate the progress of the change initiative and to invite feedback from staff. At this gathering, staff will be encouraged to openly discuss what they think is and isn't working with the change initiative. Subsequent feedback parties will be planned throughout the change initiative.

STEP 10 Revise your action and maintenance plans based on evaluation results.

1. How will you incorporate evaluation feedback into your plans?

Because staff wanted opportunities to learn more about current research, each staff person will be encouraged to spend an hour a week researching current information on the Internet and/or in the library. They will share this information at interdisciplinary team meetings.

A new team will also be coordinated to regularly distribute findings from the field through memos, newsletters and articles. A computer will be made available for this team to use in accessing information on the Internet during working hours.

Clients said they would like a way to voice their opinions confidentially in writing on a regular basis. Therefore, a questionnaire, pen and comment box will be provided in the client hallway. To encourage continued feedback from clients, a poster highlighting the changes made within the agency based on their feedback will be featured above the comment box.

2. How will you address resistance to the change initiative?

While the majority of staff are willing to participate in the change initiative, some steps are needed to encourage late adopters. These staff members have been invited to personally meet with the administrator and two opinion leaders to discuss their position. By including them in the change process directly, it is anticipated that they will become more aware of the problem and will be more willing to adopt the change. Opinion leaders will encourage their participation and feedback more often, and will take this feedback to taskforce meetings so it can be reviewed and possibly used to modify the change plan if necessary.

CONCLUDING
THOUGHT

We believe *The Change Book* will give you something "to hold on to" and will guide you as you move from the "old ways." Have confidence that the *Principles, Steps, Strategies* and *Activities* described here will lead you to new practices that will ultimately transform the lives of the people you serve. Change is hard work, but armed with these tools, you will succeed if you persevere.

"It's not so much that we're afraid of change or so in love with the old ways, but it's that place in between that we fear. . . . It's like being between trapezes. It's Linus when his blanket is in the dryer. There's nothing to hold on to."

- Marilyn Ferguson,
Author, Philosopher

EPILOGUE

By Barry Brown, PhD
University of North Carolina at Wilmington

Increasingly, if belatedly, there is a recognition of the need to commit to a program of technology transfer to allow the advances of drug treatment research to find expression in clinical practice. For years our focus has been on the generation of knowledge, with the onus for converting that knowledge to program activity, left to the service provider. This places providers in the unenviable position of having to sift through the wealth of available journals, books and government publications, and in the impossible position of teasing out the information needed to increase their own effectiveness.

Indeed, there is now an understanding that converting research findings to treatment practice is basically a task of organizational change. Even in a climate where service providers are earnestly concerned with making the best possible treatment available, we must recognize that they have developed practices they see as effective and with which they have become comfortable. Those providers work in organizations that have developed ways of operating that are seen as appropriate and have become routine. For these reasons the initiatives described in *The Change Book* represent a particularly important contribution to the successful achievement of technology transfer. There is not only a recognition of the issues involved in producing the organizational and individual change essential to technology transfer, but there is also a detailed description of the *Steps* needed to achieve that change.

It is important to emphasize that we already possess the skills and experience needed to transfer treatment research findings to treatment program practice. However, service providers want assurance that there is evidence for the effectiveness of new interventions, that these interventions give promise of adding significantly to their capacity to serve clients, and that the necessary resources to implement these interventions are available. Where these issues are addressed, technology transfer efforts can succeed.

While the success of existing technology transfer *Strategies* should give further impetus to our *Activities*, that success should not limit our efforts to refine those *Strategies* and make them more effective, or to develop and test additional technology transfer initiatives. In particular, electronic media will offer increasing opportunities for communicating information and sharing research findings. Our challenge is to understand the role electronic media can play in facilitating change given our findings regarding the importance of interpersonal strategies.

In brief, technology transfer initiatives should themselves be the subject of evaluative study allowing us to have available the best and most appropriate *Strategies* to create change for different situations and audiences. In this climate, the Substance Abuse and Mental Health Services Administration (SAMHSA) and the Addiction Technology Transfer Center (ATTC) Network play a vital role. SAMHSA and the ATTC Network are uniquely positioned between the service providing and research communities. Thus, not only are they critical to facilitating the process of technology transfer, but they can also provide direction and support to the development and study of new *Strategies* to achieve technology transfer.

The process of technology transfer also affords an opportunity to receive feedback from service providers regarding treatment issues that can lend themselves to research. The bridge between research and practice need not carry traffic in only one direction. Learning from service providers the areas where more effective treatment approaches are needed, permits researchers to be more capable of conducting research in areas of consequence. This allows both providers and researchers a capacity for even greater accomplishment.

Through the use of established change *Strategies*, technology transfer provides the means to increase treatment effectiveness by bringing research to the service delivery community. It also brings the concerns of service providers to the attention of researchers. In creating the means for the service delivery and the research communities to join forces more effectively, there is greater assurance that both will be properly responsive to the needs of the client community.

ENDNOTES

1. Webster's Third New International Dictionary of the English Language Unabridged. (1971). Chicago: Merriam & Co.

2. Ibid.

3. Ibid.

4. Institute of Medicine. (1999). New partnerships for a changing environment: Why drug and alcohol treatment providers and researchers need to collaborate. Washington, DC: National Academy Press.

5. Brown, B. S. (2000). From research to practice - The bridge is out and the water's rising. In J. A. Levy, R. C. Stephens, & D. C. McBride (Eds.), Emergent issues in the field of drug abuse: Advances in medical sociology (Vol. 7, pp. 345-365). Stanford, CT: JAI Press.

6. McCarty, D. (1999, November). Treatment innovations: Implementation strategies for practitioners, organizations and systems. Presented at the Addiction Technology Transfer Center Technology Transfer Symposium, Alexandria, VA.

7. Valente, T. W. (1999, November). Models and methods for accelerating technology transfer. Presented at the Addiction Technology Transfer Center Technology Transfer Symposium, Alexandria, VA.

8. Velasquez, M. M. (1999, November). The application of the Transtheoretical Model of Change to addiction technology transfer. Presented at the Addiction Technology Transfer Center Technology Transfer Symposium, Alexandria, VA.

9. DiMaggio, P., & Unseem, M. (1979). Decentralized applied research: Factors affecting the use of audience research by arts organizations. Journal of Applied Behavioral Science, 15, 79-94.

10. Boyer, J. F., & Langbein, L. I. (1991). Factors influencing the use of health evaluation research in Congress. Evaluation Review, 15, 507-532.

11. Argarawala-Rogers, R. (1977). Why is evaluation research not utilized? In M. Guttentag (Ed.), Evaluation Studies Review Annual (Vol. 11). Beverly Hills, CA: Sage.

12. Boyer, J. F., & Langbein, L. I. (1991).

13. Valente, T. W. (2000, April). Knowledge application strategies: Knowing your audience and designing the best strategy. Presented at the Center for Substance Abuse Treatment Practice/Research Collaborative Grantee Meeting, New Orleans, LA.

14. Brown, B. S. (2000).

15. Prochaska, J. O., Norcross, J. C., & DiClemente, C. C. (1994). Changing for good: A revolutionary six-stage program for overcoming bad habits and moving your life positively forward. New York: Avon Books.

OTHER RESOURCES

The following list of resources is new to this second edition of The Change Book.

Altman, D. G. (1995). Sustaining interventions in community systems: On the relationship between researchers and communities. *Health Psychology, 14(6),* 526-536.

Anderson, R. A., Issel, L. M., & McDaniel, R. R. (2003). Nursing homes as complex adaptive systems. *Nursing Research, 52,* 12-21.

Andrzejewski, M. E., Kirby, K. C., Morral, A. R., & Iguchi, M. Y. (2001). Technology transfer through performance management: The effects of graphical feedback and positive reinforcement on drug treatment counselors' behavior. *Drug and Alcohol Dependence, 63,* 179-186.

Axelrod, R., & Cohen, M. D. (2000). *Harnessing complexity: Organizational implications of a scientific frontier.* New York: Simon & Schuster.

Backer, T. E. (2000). The failure of success: Challenges of disseminating effective substance abuse prevention programs. *Journal of Community Psychology, 28(3),* 363-373.

Bero, L. A., Grilli, R., Grimshaw, J. M., Harvey, E., Oxman, A. D., & Thomson, M. A. (1998). Closing the gap between research and practice: An overview of systematic reviews of interventions to promote the implementation of research findings. *British Medical Journal, 317,* 465-468.

Bosworth, K., Gingiss, P. M., Potthoff, S., & Roberts-Gray, C. (1999). A Bayesian model to predict the success of the implementation of health and education innovations in school-centered programs. *Evaluation and Program Planning, 22,* 1-11.

Brown, B. S. (2000). From research to practice: The bridge is out and the water is rising. *Advances in Medical Sociology, 7,* 345-365.

Browning, L. D. (1992). Lists and stories as organizational communication. *Communication Theory, 2(4),* 281-302.

Cunningham, J. A., Martin, G. W., Coates, L., Here, M. A., Turner, B. J., & Cordingley, J. (2000). Disseminating a treatment program to outpatient addiction treatment agencies in Ontario. *Science Communication, 22(2),* 154-172.

Davis, D., Evans, M., Jadad, A., Perrier, L., Rath, D., Ryan, D., Sibbald, G., Straus, S., Rappolt, S., Wowk, M., & Zwarenstein, M. (2003). Learning in practice: The case for knowledge translation: Shortening the journey from evidence to effect. *British Medical Journal, 327,* 33-35.

Diamond, M. A., (1996). Innovation and diffusion of technology: A human process. *Consulting Psychology Journal: Practice and Research, 48(4),* 221-229.

Dietrich, A. J., Woodruff, C. B., & Carney, P. A. (1994). Changing office routines to enhance preventive care: The preventive GAPS Approach. *Archives of Family Medicine, 3*, 176-183.

Enderby, J.E., & Phelan, D.R. (1994). Action learning groups as the foundation for cultural change. *The Quality Magazine, 3*, 42-49.

Ferrence, R. (2001). Diffusion theory and drug use. *Addiction, 96(1)*, 165-174.

Forman, R. F., Bovasso, G., Woody, G. (2001). Staff beliefs about addiction treatment. *Journal of Substance Abuse Treatment, 21*, 1-9.

Glasgow, R. E., Bull, S. S., Gillette, C., Klesges, L. M., & Dzewaltowski, D. A. (2002). Behavior change intervention research in health care settings: A review of recent reports with emphasis on external validity. *American Journal of Preventive Medicine, 23*, 62-69.

Glasgow, R. E., Lichenstein, E., & Marcus, A. C. (2003). Why don't we see more translation of health promotion research to practice? Rethinking the efficacy-to-effectiveness transition. *Public Health Matters, 93(8)*, 1261-1267.

Green, L. W., Gottlieb, N. H., & Parcel, G. S. (1991). Diffusion theory extended and applied. *Advances in Health Education and Promotion, 3*, 91-117.

Gustafson, D. H., & Hundt, A. S. (1995). Findings of innovation research applied to quality management principles for health care. *Health Care Management Review, 20(2)*, 16-33.

Jensen, P. S., Hoagwood, K., & Trickett, E. J. (1999). Ivory towers or earthen trenches: Community collaborations to foster real-world research. *Applied Developmental Science, 3(4)*, 206-212.

Jensen, P. S., Virello, B., Bhatara, V., Hoagwood, K., & Feil., M. (1999). Psychoactive medication prescribing practices for U.S. children: Gaps between research and clinical practice. *Journal of the American Academy of Child and Adolescent Psychiatry, 38*, 557-565.

Kavanagh, K. H. (1995). Collaboration and diversity in technology transfer. In T. Backer, S. David, & G. Saucy, (Eds.), *Reviewing the behavioral science base on technology transfer* (pp. 42-64), National Institute on Drug Abuse Monograph Series No. 155. Rockville, MD: U.S. Government Printing Office.

McCarty, D., Rieckmann, T., Green, C., Gallon, S., & Knudsen, J. (2004). Training rural practitioners to use buprenorphine: Using *The Change Book* to facilitate technology transfer. *Journal of Substance Abuse Treatment, 26*, 203-208.

Miller, W. L., Crabtree, B. F., McDaniel, R., & Stange, K. (1998). Understanding change in primary care practice using Complexity Theory. *The Journal of Family Practice, 46(5)*, 369-376.

Miller, W. L., McDaniel, R. R., Crabtree, B. F., Stange, K. C. (2001). Practice jazz: Understanding variation in family practices using complexity science. *The Journal of Family Practice, 50(10)*, 872-880.

Miller, W. R., & Mount, K. A. (2001). A small study of training in motivational interviewing: Does one workshop change clinician and client behavior? *Behavioural and Cognitive Psychotherapy, 29*, 457-471.

Milne, E., Westerman, D., & Hanner, S. (2002). Can a "relapse prevention" module facilitate the transfer of training? *Behavioural and Cognitive Psychotherapy, 30*, 361-364.

Murphy-Smith, M., Meyer, B., Hitt, J., Taylor-Seehafer, M. A., & Tyler, D. O. (2004). Put prevention into practice implementation model: Translating practice into theory. *Journal of Public Health Management Practice, 10(2),* 109-115.

Rogers, E. M. (2002). The nature of technology transfer. *Science Communication, 23(3),* 323-341.

Rosenchek, R. A. (2001). Organizational process: A missing link between research and practice. *Psychiatric Services, 52(12),* 1607-1612.

Sechrest, L., Backer, T. E., Rogers, E. M., Campbell, T. F., & Grady, M. L. (Eds.) (1994). *Effective dissemination of clinical and health information.* Rockville, MD: Agency for Health Care Policy & Research.

Sorenson, J. L., Hall, S. M., Loeb, P., Allen, T., Glaser, E. M., & Greenberg, P. D. (1988). Dissemination of a job seeker's workshop to drug treatment programs. *Behavior Therapy, 19,* 143-155.

Sorenson, J. L., Rawson, R. A., Guydish, J., & Zweben, J. (2003). *Drug abuse treatment through collaboration: Practice and research partnerships that work.* Washington, D.C.: American Psychological Association.

Thompson, R. S., Taplin, S., McAfee, T. A., Mandelson, M. T., & Smith, A. E. (1995). Primary and secondary prevention services in clinical practice: Twenty years experience in development, implementation, and evaluation. *Journal of the American Medical Association, 273(14),* 1130-1135.

Valente, T. W., & Davis, R. L. (1999). Accelerating the diffusion of innovations using opinion leaders. *Annals of the American Academy of Political and Social Science, 566,* 55-67.

Wagner, E. H., Austin, B. T., Davis, C., Hindmarsh, M., Schaefer, J., & Bonomi, A. (2001). Improving chronic illness care: Translating evidence into action. *Health Affairs, 20,* 64-78.

Weber, V., & Joshi, M. S. (2000). Effecting and leading change in health care organizations. *The Joint Commission Journal on Quality Improvement, 26,* 388-399.

TOOLS FOR CHANGE

The following list of resources is new to this second edition of **The Change Book.**

Corrigan, P. W., Artarin, K. W. & Pramana, W. Y. (1992). Staff perceptions of behavior therapy at a at a psychiatric hospital. *Behavior Modification*, No. 16, 132-134.

Corrigan, P. W., Holmes, E. P., Luchins, D., Parks, J., Basit, A., Dehaney, E., & Kayton-Weinberg, D. (1994). Setting up inpatient behavioural treatment programmes: The staff needs assessment. *Behavioural Interventions*, 9, 1-12.

Evans, A., Gustafson, D., Ketley, D., Maher, L., & McManus, L. (2002). *British Health Services Sustainability Model.* Leicister, England: British National Health Service Modernization Agency.

Humphreys, K., Greenbaum, M. A., Noke, J. M., Finney, J. W. (Mar 1996). Reliability, validity, and normative data for a short version of the Understanding of Alcoholism Scale. *Psychology of Addictive Behaviors. 10(1)*, 38-44.

Miller, W., Hedrick, K., & Orlofsky, D. (1991). The helpful responses questionnaire: A procedure for measuring therapeutic empathy. *Journal of Clinical Psychology, 47*, 444-448.

Moos, R. H. (1981). *Work Environment Scale Manual* (2nd ed.). Palo Alto, CA: Consulting Psychologists Press.

Moyers, T. and Miller, W. (1993). Therapist's conceptualizations of alcoholism: Measurement and implications for treatment decisions. *Psychology of Addictive Behaviors, 7(4)*, 238-245.

Simpson, D. D. (2001). *Core set of TCU forms.* Includes: 1) Organizational Readiness for Change, 2) Program Training Needs, and 3) Client Evaluation of Self and Treatment. Fort Worth: Texas Christian University, Institute of Behavioral Research. [Online]. Available: www.ibr.tcu.edu/pubs/datacoll/datacoll.html#TCUCommunity.

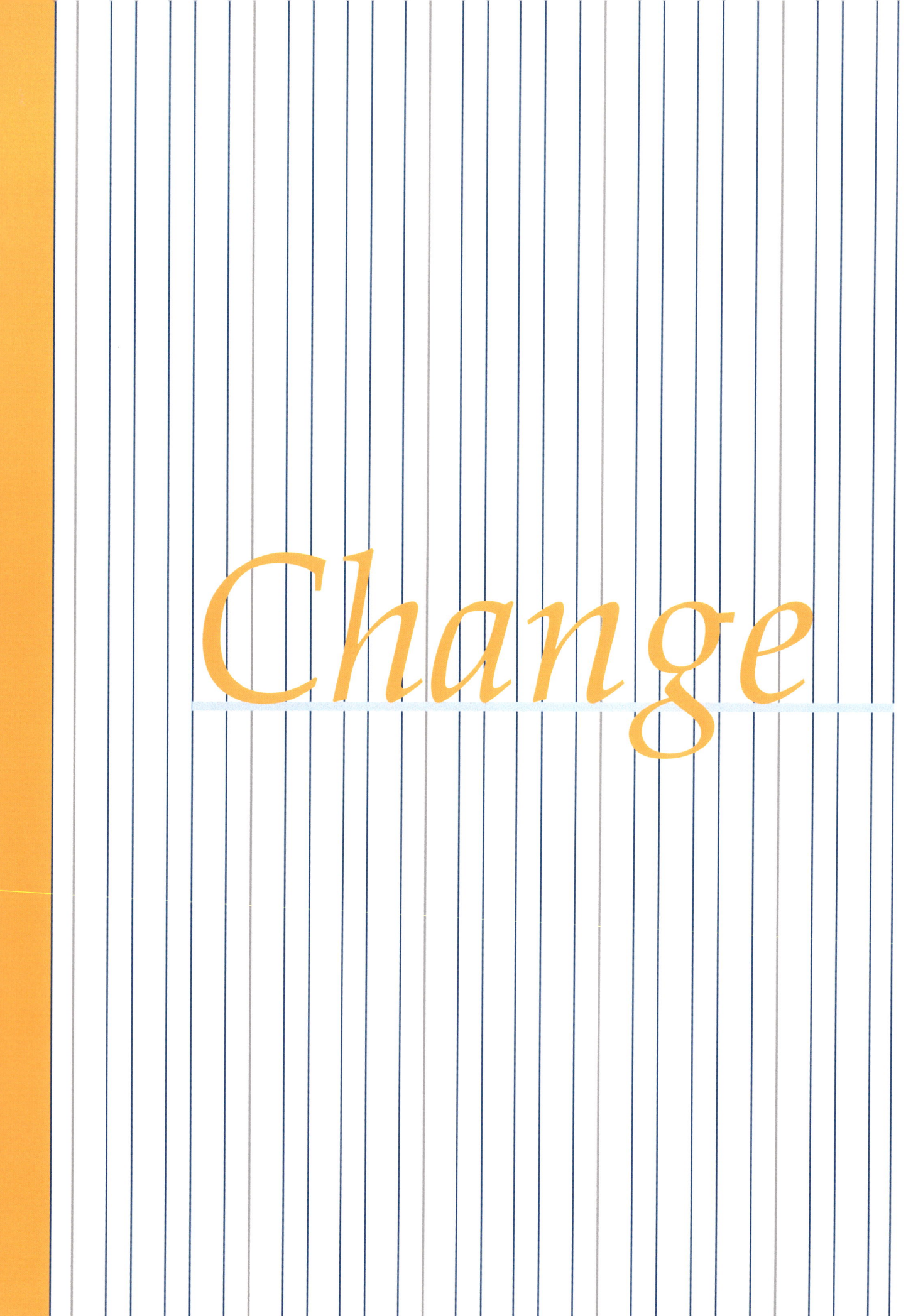

APPENDIX

SYNTHESIS OF PRESENTATIONS AT THE
ATTC TECHNOLOGY TRANSFER SYMPOSIUM

In November 1999, the ATTC Network hosted a Technology Transfer Symposium. This Symposium was designed to keep staff across the country informed about what current research indicates works and doesn't work in technology transfer. Guests and presenters at the Symposium included:

JON GOLD, presented an overview of CSAT's technology transfer initiatives and the role CSAT plays in the adoption of best practices within treatment systems.

THOMAS VALENTE, PHD, summarized current research, highlighted effective technology transfer *Principles* and explained the concept of utilizing opinion leaders in technology transfer.

DENNIS MCCARTY, PHD, discussed the application of technology transfer strategies to individuals, organizations and systems, utilizing the example of the slow adoption rate of medications in the treatment of substance use disorders as a case in point.

MARY MARDEN VELASQUEZ, PHD, discussed the stages of change, readiness for change and the complexity of technology transfer at various stages of change.

The researchers and professionals presenting at the Symposium suggested a variety of approaches and Strategies for increasing the effectiveness of technology transfer. The pages that follow are a synthesis of the approaches and *Strategies* that have the broadest application, and support the purposes of this guide.

JON GOLD
CENTER FOR SUBSTANCE ABUSE TREATMENT'S KNOWLEDGE APPLICATION PROGRAM (KAP)

Gold's presentation at the 1999 ATTC Technology Transfer Symposium provided an overview of the Center for Substance Abuse Treatment's (CSAT) technology transfer initiatives and the role CSAT plays in the adoption of best practices within treatment systems.

Knowledge, Development, Application Programs (KDAs), the ATTC, research and evaluation contracts, single state agencies and a number of other federal agencies have contributed to a vast knowledge base at CSAT about the substance abuse treatment field. A new CSAT initiative designed to apply this knowledge is called the Knowledge Application Program (KAP). KAP's goals are to:

1. Ensure coordination and strengthen collaboration among CSAT's various knowledge application initiatives.
2. Capture information available in CSAT's knowledge base and make it accessible to target audiences, using effective formats and dissemination channels.
3. Revise or repackage current CSAT products so they are relevant and applicable to culturally diverse audiences.
4. Use state-of-the-art knowledge application methods to increase the adoption of research-based, best practice guidelines.
5. Strengthen and broaden CSAT's dissemination channels.
6. Increase awareness of CSAT products.

Potential users of this information include substance use disorder treatment providers, state and federal agencies, consumers, families and CSAT staff. KAP will use multiple communication channels to bring information to these audiences.

DENNIS MCCARTY, PHD
TREATMENT INNOVATIONS: IMPLEMENTATION STRATEGIES FOR PRACTITIONERS, ORGANIZATIONS AND SYSTEMS

Increasing the effectiveness of technology transfer efforts will require change at a variety of levels within the overall alcohol and other drug treatment system – including patients, practitioners and programs. There will be barriers to change at each level and different strategies required if practices within each level are to change. According to Dennis McCarty, PhD, in his presentation at the ATTC Technology Transfer Symposium, the challenge is finding strategies to promote the adoption of new technology at the individual level, the practitioner/clinical level and the program/organizational level.

STRATEGIES FOR AFFECTING ATTITUDE CHANGE AND ADOPTION OF INNOVATIONS AT THE CLIENT/PATIENT LEVEL:

1. Provide evidence of how a practice or innovation works.
2. Educate the client/patient about the innovation.
3. Refer to effective technologies in other areas or fields.
4. Utilize advertising and marketing plans.

STRATEGIES FOR AFFECTING ATTITUDE CHANGE AND ADOPTION OF INNOVATIONS AT THE PRACTITIONER/CLINICAL LEVEL:

1. Provide evidence of how a practice or innovation works.
2. Educate the practitioner about the innovation.
3. Refer to effective technologies in other areas or fields.
4. Provide incentives for clinicians to use an innovation (peer support, financial incentives, outcomes monitoring).
5. Identify early adopters and allow them to model.
6. Gain single state agency involvement in the adoption of an innovation.
7. Utilize a multifaceted approach to behavior change.
8. Utilize advertising and marketing plans.

STRATEGIES FOR AFFECTING ATTITUDE CHANGE AND ADOPTION OF INNOVATIONS AT THE PROGRAM/ORGANIZATION LEVEL:

1. Secure single state agency support and/or funding.
2. Educate programs/organizations that financing is important but should not be the "end all."
3. Provide responses to the concerns or barriers perceived by the programs/organizations.
4. Develop training and diffusion strategies specifically for small stand-alone treatment programs and those with staff in recovery and/or without degrees.

OTHER FACTORS THAT CAN AFFECT THE ADOPTION OF INNOVATIONS

> The size of a program/organization - larger programs are more likely to adopt new innovations.
> Type of work setting – community-based treatment program, medical or mental health center program, freestanding clinics.
> Staff composition – number of staff in recovery, staff with licenses, staff with master's or doctoral degrees.
> The champion of an innovation or practice affects its adoption.
> Different learning styles require different strategies for acceptance and adoption of new innovations.

THOMAS VALENTE, PHD
MODELS AND METHODS FOR ACCELERATING TECHNOLOGY TRANSFER

Thomas Valente, PhD, in his presentation at the 1999 ATTC Technology Transfer Symposium, points out that technology transfer takes time and there are many different players in the process (e.g., researchers, government agencies and treatment organizations). In the transfer of technology it is important to observe and manage the agenda setting process (which determines what is studied), dissemination, diffusion and utilization processes, and feed the results back into the system. Valente also reminds us that the introduction of practice standards or new innovations can have both positive and negative dimensions:

POSITIVE
> Designed to improve practice
> Facilitates interchange
> Standardizes protocols
> Facilitates analysis

NEGATIVE
> Upsets existing procedures
> Destabilizes structures
> Threatens the status quo

There is often a trade-off between the *impact* or effect of a message on the intended audience and the *reach* or percent of the intended audience exposed to a message. The most effective strategies for diffusion of innovations reach a large number of people and create a lot of change. Diffusion of innovations explains how new ideas and practices spread:

> People vary in their innovativeness and adopt at different stages (some early, some later).
> Perceived characteristics of the innovation influence adoption.
> People adopt in stages: awareness, learning, attitude, trial and routine use.

Research in other fields can be applied to the transfer of technology in the field of substance use disorders. For instance, lessons learned from communications research indicate *what works* in technology transfer and diffusion of innovations.

1. Utilize active strategies rather than passive ones.
2. Multifaceted interventions are more effective than single-faceted ones.
3. Continuous, rather than static efforts or one-shot programs are more effective.

Recent research also indicates that *opinion leaders* are particularly influential in the adoption of new innovations. Opinion leaders are not necessarily the same as "leaders" on an organizational chart. Rather, they are selected by the rest of the group. Opinion leaders tend to be conservative, espouse dominant norms, and they tend to wait to see where a group is going before adopting something new. Evidence suggests that opinion leaders can make dramatic differences in the final outcome/adoption of a practice and how widely it is actually adopted.

FACTORS THAT INFLUENCE BEHAVIOR

> Behavior of colleagues and peers.
> Perceptions of approval by peers.
> Active encouragement by peers.
> Perceived support by opinion leaders for a new behavior.
> Individuals learn from their interaction with role models and colleagues.

Utilization of findings from other studies and fields can help the substance use disorder field accelerate the process of adoption/diffusion of new practices or knowledge. Use of opinion leaders is not the only strategy for accelerating the adoption process, but it is one strategy that can be explored.

METHODS FOR IDENTIFYING OPINION LEADERS AND THE
ADVANTAGES/DISADVANTAGES OF EACH

Self-selection - Staff requests volunteers in-person or via mass media. Those who volunteer are selected.
Advantage: Easy to implement   *Disadvantage:* Not a valid measure

Self-identification - Surveys are administered to the sample, and questions measuring leadership are included. Those scoring highest on leadership scales are selected.
Advantage: Easy to implement   *Disadvantage:* Not a valid measure

Staff-selected - Program implementers select leaders from those whom they know.
Advantage: Easy to implement   *Disadvantage:* Dependent on staff's ability

Positional approach - Persons occupying leadership positions such as clergy, elected officials, media, business elites and so on are selected.
Advantage: Easy to implement   *Disadvantage:* Positional leaders may not be leaders for the community.

Judge's ratings - Persons who are knowledgeable identify leaders to be selected.
Advantage: Easy to implement   *Disadvantage:* Dependent on the selection of raters and their ability to rate.

Expert identification - Trained ethnographers study communities to select leaders.
Advantage: Implementation can   *Disadvantage:* Dependent on abilities of experts (ethnographers).
be done in many settings.

Snowball method - Index cases provide nominations of leaders who are in turn interviewed until no new leaders are identified.
Advantage: Implementation can be done   *Disadvantage:* Results are dependent on the representatives of the index
in many settings; provides some   cases, and it can take considerable time to trace individuals who are
measure of the network.   nominated.

Sample sociometric - Randomly selected respondents nominate leaders and those receiving frequent nominations are selected.
Advantage: Implementation can   *Disadvantage:* Results are dependent on representativeness of sample,
be done in many settings; provides   and may be restricted to communities with less than 1,000 members.
some measure of the network, only
requires one measurement.

Sociometric - All or most respondents interviewed and those receiving frequent nominations are selected.
Advantage: Network of community   *Disadvantage:* Time consuming and expensive to interview everyone.
can be mapped and other centrality   Results are dependent on the representativeness of the sample and may be
techniques used to locate opinion   restricted to relatively small communities (i.e. less than 1,000 members).
leaders; only requires one measure;
has highest validity and reliability.

MARY MARDEN VELASQUEZ, PHD
THE APPLICATION OF THE TRANSTHEORETICAL MODEL OF CHANGE TO ADDICTION TECHNOLOGY TRANSFER

The Transtheoretical Model of Change has been applied to drug problems, smoking cessation, exercise adoption, diet, condom use, health promotion and more. The model is identified as appropriate to addiction technology transfer because it is intuitive and has broad application. The model recognizes that organizations and individuals are in various stages of change when presented with a technology transfer initiative. As emphasized in McCarty's presentation at the Symposium, there are different levels or targets for behavior change, which include the research community, policy makers, supervisors and field clinicians.

Velasquez highlighted the importance of "marketing" any innovation or practice by taking into account the primary audience or "target" for change and the various stages of change the target audience might be in. Assessing the stage of change (of an agency) is as important as assessing an individual client's stage of change. Strategies employed in technology transfer will not be effective in bringing about change in practice or use of new innovations if the "message" is ahead of an entity's stage of change.

STAGES OF CHANGE

PRECONTEMPLATION: People and organizations in this stage tend to be content with the status quo. If things are working why change? Some communication strategies that may help move these people toward change include:

1. Consciousness-raising.
2. Use of different mediums for dissemination of technology and multiple attempts.
3. Conduct needs assessments.
4. Recognize that people and organizations are at this stage for different reasons.
5. Assess decisional balance and elicit the good things versus the not-so-good things about change.

CONTEMPLATION: People and organizations thinking about change can be overwhelmed with too much information. They need just enough to make them interested.

1. Provide "tastes" of the topic to build interest.
2. Provide evidence for effectiveness of a new technology (not just statistics).
3. Probe for their issues of concern.
4. Build self-efficacy: a person's belief in his or her ability to carry out or succeed with a specific task. Treatment professionals need this as much as clients do.
5. "Tip" the decisional balance. Help to identify more "pros" than "cons" to help move people toward change.

PREPARATION: People and organizations are getting ready to make a change. Movement to the "action" stage of change is not smooth, and preparation becomes an important step.

1. Be sure the language and format of a given technology are clear.
2. Assist in developing a change plan.
   a. How can the technology best be replicated in a particular setting?
   b. Are there site-specific barriers to implementation?

ACTION: People and organizations are actively changing. It is important to support people and organizations in change. We often focus on getting people and organizations to buy into change and withdraw support once the action stage is reached.

1. Provide information in a "user-friendly" fashion.
2. Encourage questions and problem-solving.
3. Have frequent interpersonal contact – mentoring during this stage is important.
4. Provide ongoing monitoring.
5. Offer nonthreatening feedback.

MAINTENANCE: Continue the behavior change. It is important to focus on maintenance of a new behavior so people and organizations follow through and don't just move on to the next "innovation."

1. Continue communication (updates, newsletters, Web sites, listservs, telephone trees).
2. Continue interpersonal contacts.
3. Encourage communication and problem-solving.

It is important to remember that people and organizations go through stages of change several times. They can experience "relapse."

ATTC IN BRIEF

Serving the Field
Since 1993

Building on a rich history, the Addiction Technology Transfer Center (ATTC) Network is dedicated to identifying and advancing opportunities for improving addictions treatment and recovery services. Our vision is to unify science, education and services to transform the lives of individuals and families affected by alcohol and other drug addiction.

The ATTC Network undertakes a broad range of initiatives that respond to emerging needs and issues in the treatment field. The Network is funded by the Substance Abuse and Mental Health Services Administration (SAMHSA) to upgrade the skills of existing treatment practitioners and other health professionals, and to disseminate the latest scientific findings to the treatment community. We expend those resources to create a multitude of products and services that are timely and relevant to the many disciplines represented by the addiction treatment workforce.

Serving the 50 U.S. States, the District of Columbia, Puerto Rico, the U.S. Virgin Islands and the Pacific Islands, the ATTC Network operates as 14 individual Regional Centers and a National Office. At the regional level, individual Regional Centers focus primarily on meeting the unique needs in their areas while also supporting national initiatives. The National Office leads the Network in implementing national initiatives and concurrently supports and promotes individual regional efforts.

Together we take a unified approach in delivering cutting-edge knowledge and skills that develop a powerful workforce . . . a workforce that has the potential to transform lives.

For more resources on technology transfer, visit:
ATTCnetwork.org/techtransfer

The Change Book *Workbook*

A Companion to *The Change Book:*
A Blueprint for Technology Transfer

ATTCnetwork.org

Unifying science, education
and services to transform lives.

Addiction Technology Transfer Center Network
Funded by Substance Abuse and Mental Health Services Administration

Workbook

TABLE OF CONTENTS

THE CHANGE BOOK RECAP   2

TEN *STEPS* to Effective Technology Transfer   3

PRINCIPLES of Effective Technology Transfer, Organizing Your Team, Addressing Resistance   4

STRATEGIES to Use for Each Stage of Change   5

STRATEGIES to Use With Multiple Levels of an Organization   6

ACTIVITIES IDEA LIST   7

BARRIERS TO CHANGE   8

WORKBOOK: STEP 1   11

WORKBOOK: STEP 2   12

WORKBOOK: STEP 3   14

WORKBOOK: STEP 4   15

WORKBOOK: STEP 5   17

WORKBOOK: STEP 6   19

WORKBOOK: STEP 7   20

WORKBOOK: STEP 8   23

WORKBOOK: STEP 9   23

WORKBOOK: STEP 10   25

Share this Workbook

Our hope is that you can use this workbook for multiple change initiatives and will share the workbook pages with all members of your change team.

RECAP

THE CHANGE BOOK RECAP

Now that you've spent some time immersed in the basics of technology transfer, it is time to create your own blueprint for change. The *Principles, Strategies,* and *Activities* outlined in *The Change Book: A Blueprint for Technology Transfer (The Change Book)* are all recapped on the following pages to help you create a winning change initiative.

A workbook is also included with space for you to apply the *Steps* (and sample questions that follow each *Step*) to your situation. We hope you'll make this workbook a starting place in creating a change plan that meets the needs of your organization and team. Remember the questions under each *Step* were based on the case study discussed in *The Change Book.* These questions may be appropriate for your own change plan, or you may need to add, delete or adapt the questions under each *Step* based on your own needs. To increase the likelihood of developing a successful change initiative, however, answer all the questions you include as completely as possible.

Also remember that some of the *Steps* may be worked simultaneously, or the order of the *Steps* may be changed to fit your needs. For your plan to succeed, however, *Steps* 1-7 should be completed before you implement your change plan in *Step* 8.

TIPS

ADD OR DELETE QUESTIONS UNDER EACH STEP BASED ON YOUR NEEDS.

ANSWER ALL QUESTIONS UNDER EACH STEP AS COMPLETELY AS POSSIBLE.

WORK *STEPS* 1-7 PRIOR TO IMPLEMENTING YOUR CHANGE INITIATIVE.

TEN STEPS
to Effective Technology Transfer

To change your agency or system from what it is now into what you want it to be, you'll need a blueprint to guide you. The Steps that follow provide a starting place. Some of the Steps may be worked simultaneously, or the order of the Steps may be changed to fit your needs. **For your plan to succeed, however, Steps 1-7 should be completed before you implement your change plan (Step 8).**

STEP 1
Identify the problem.

STEP 2
Organize a team for addressing the problem.

STEP 3
Identify the desired outcome.

STEP 4
Assess the organization or agency.

STEP 5
Assess the specific audience(s) to be targeted.

STEP 6
Identify the approach most likely to achieve the desired outcome.

STEP 7
Design action and maintenance plans for your change initiative.

STEP 8
Implement the action and maintenance plans for your change initiative.

STEP 9
Evaluate the progress of your change initiative.

STEP 10
Revise your action and maintenance plans based on evaluation results.

PRINCIPLES

of Effective Technology Transfer, Organizing Your Team, Addressing Resistance

OF EFFECTIVE TECHNOLOGY TRANSER

> **RELEVANT**
> *The technology in question must have obvious, practical application.*

> **TIMELY**
> *Recipients must acknowledge the need for this technology now or in the very near future.*

> **CLEAR**
> *The language and process used to transfer the technology must be easily understood by the target audience.*

> **CREDIBLE**
> *The target audience must have confidence in the proponents/sources of the technology.*

> **MULTIFACETED**
> *Technology transfer will require a variety of Activities and formats suited to the various targets of change.*

> **CONTINUOUS**
> *The new behavior must be continually reinforced at all levels until it becomes standard and then is maintained as such.*

> **BI-DIRECTIONAL**
> *From the beginning of the change initiative, individuals targeted for change must be given opportunities to communicate directly with plan implementers.*

ORGANIZING YOUR TEAM

We encourage you to use a team approach from beginning to end with any change initiative. It is important to build your team with people from all levels of your agency. Your team's size will depend on the size of your organization and the particular change initiative you are imple¬menting. Include opinion leaders and early adopters in your team.

ADDRESSING RESISTANCE

Resistance at all levels of the organization should be expected and will require attention. Thoroughly explain how making a change can save time, enhance skills and benefit clients.

MINIMIZING RESISTANCE
> Directly address resistance
> Listen to fears and concerns
> Discuss pros and cons openly
> Educate and communicate
> Provide incentives and rewards
> Develop realistic goals
> Celebrate small victories
> Actively listen to resistors
> Actively involve as many people as possible from the beginning
> Emphasize that feedback will shape the change process
> Use opinion leaders and early adopters for training and promotion

STRATEGIES
to Use for Each Stage of Change

PRECONTEMPLATION

1. Raise the awareness of this group about the approach under consideration.
2. Use a variety of media to disseminate information.
3. Make multiple attempts to disseminate information.
4. Conduct a needs assessment. Evaluate current practices and share results.
5. Recognize that people and organizations are at this stage of change for different reasons.
6. Assess the decisional balance and elicit conversation regarding the benefits versus the drawbacks about making a change.

CONTEMPLATION

1. Provide "tastes" of the topic to build interest.
2. Provide evidence for effectiveness of a recommended approach. Don't just provide statistics.
3. Probe the group to learn their reasons for concern.
4. Build self-efficacy: a person's belief in his or her ability to carry out or succeed with a specific task. (Treatment professionals need this as much as clients do.)
5. "Tip" the decisional balance. Help to identify more pros than cons about the recommended approach to build confidence in the initiative and move people toward change.

PREPARATION

1. Be sure the language and format of the information you disseminate are clear to your target audiences.
2. Assist in the development of a change plan.
3. Make sure the change can be adopted in your particular setting.
4. Remove any site-specific barriers to implementation.

ACTION

1. Provide information in a "user-friendly" fashion.
2. Encourage questions and problem-solving.
3. Have frequent interpersonal contact.
4. Mentoring during this stage is important.
5. Provide ongoing monitoring.
6. Offer nonthreatening feedback.

MAINTENANCE

1. Continue communication (updates, newsletters, Web sites, listservs, telephone trees).
2. Continue interpersonal contact.
3. Encourage communication and problem-solving.
4. Develop skills to maintain the behavior.

STRATEGIES
to Use with Multiple Levels of an Organization

PROGRAM/ORGANIZATION LEVEL

When addressing this level, it's important to:

1. Provide evidence of how the recommended approach works.
2. Inform agencies and organizations that although financing is important, it cannot and should not be the basis for deciding whether to address a needed change. Many changes can be made with limited time and finances.
3. Secure the tangible support (financial or other) of stakeholders and funders who have policy-making authority, such as a single state agency, grantor, board, etc.
4. Acknowledge and respond to the concerns or barriers perceived by the agency or organization.
5. Develop training and diffusion *Strategies* that are suited and will appeal to each of the target groups that makeup the organization.

PRACTICIONER/CLINICAL LEVEL

When addressing this level, it's important to:

1. Provide evidence of how the recommended approach works.
2. Educate the practitioner about the approach.
3. Refer to the effectiveness of related or parallel technologies in other areas or fields.
4. Provide incentives for clinicians to use a recommended approach (peer support, financial incentives, outcomes monitoring).
5. Identify early adopters and allow them to model the new behavior.
6. Utilize a multifaceted approach to behavior change.
7. Utilize advertising and marketing to get the word out to staff.

CLIENT PATIENT LEVEL

When addressing this level, it's important to:

1. Provide evidence of how a recommended approach works.
2. Educate the client/patient about the approach.
3. Refer to the effectiveness of related or parallel technologies in other areas or fields.
4. Utilize advertising and marketing to get the word out to clients.

For additional resources on creating change plans, visit: ATTCnetwork.org

ACTIVITIES IDEA LIST

Keep in mind that effective technology transfer is not one-dimensional and therefore cannot include only one activity. Some Activities can be implemented agency-wide, others will be used one-on-one with individuals.

ADMINISTRATIVE/ STRUCTURAL ACTIVITIES

> Develop strategic plans
> Implement legal and funding mandates
> Implement policy changes
> Provide on-site technical assistance
> Provide rewards/incentives for change (intrinsic or extrinsic)

PERSON-TO-PERSON ACTIVITIES

> Conduct mentoring
> Encourage peer-to-peer coaching
> Provide clinical supervision
> Use early adopter influence
> Use opinion leader influence
> Utilize role playing

EVALUATION ACTIVITIES

> Collect baseline data
> Conduct needs assessments
> Conduct outcome/impact studies
> Conduct process evaluation
> Develop reports

EDUCATIONAL ACTIVITIES

> College courses
> Conference workshops
> Education groups within your agency
> Lectures
> Online courses
> Professional meetings
> Quizzes and examinations
> Self-directed learning packages
> Short training courses (1-5 days/topic specific)
> Workshop training sessions

INFORMATION DISSEMINATION ACTIVITIES

> Ads and public service announcements
> Audiotapes
> Books/manuals
> Curriculum packages
> E-zines (online magazines)
> Fact sheets
> Government publications
> Internal reports with results/ accomplishments
> Memos
> Newsletter articles
> Posters
> Press releases
> Professional journal articles
> Promotional flyers
> Teleconferences
> Video instruction
> Web sites

BARRIERS TO CHANGE

SYSTEM STRUCTURE

THE BARRIER: *Federal, state and local government entities and individual agencies charged with responsibility for the prevention and treatment of substance use disorders are fragmented, don't communicate, and often work at cross-purposes.*

THE OPPORTUNITY: *These systems provide fertile ground for change efforts such as cross-training initiatives that improve client outcomes and increase cross-system collaborations.*

LOCAL BARRIERS:

_____

_____

_____

_____

LOCAL OPPORTUNITIES:

_____

_____

_____

_____

THE POLICY MAKERS

THE BARRIER: *Community-based treatment agencies often receive federal, state, health insurance and private funds. These funding sources may not support or may be in conflict about funding innovative research-based treatment methods. Public or payor policies may not support the application of new scientific discoveries, especially when they challenge established and familiar practices and beliefs.*

THE OPPORTUNITY: *Community organizations collaborating with researchers are ideally positioned to educate policy makers about the efficacy of research-based methodologies.*

LOCAL BARRIERS:

_____

_____

_____

_____

LOCAL OPPORTUNITIES:

_____

_____

_____

_____

THE RESEARCH COMMUNITY

THE BARRIER: *Most scientific research is rewarded by publication in professional journals. These journals are often not available to the clinical practice community because journal subscriptions can be costly and tend to be written for scientific audiences. Formal training for clinicians seldom includes practical lessons in using research literature to improve and change practice. Even when available, research reports typically do not meet the practical needs of frontline staff wanting to apply new research findings.*[11]

THE OPPORTUNITY: *Increase the occasions for dialogue between researchers and treatment organizations to heighten awareness about the need for and benefit of collaborative relationships between the two groups.*

LOCAL BARRIERS:

_____

_____

_____

_____

LOCAL OPPORTUNITIES:

_____

_____

_____

_____

AGENCY STAFF

THE BARRIER: *Staff can be resistant to change due to many factors including: a lack of understanding the new information, a lack of incentive for change, competing priorities, funding limitations, fear of failure and a general fear of change.*

THE OPPORTUNITY: *When properly addressed, opportunities may include cross-training of staff, open dialogue between clinicians and administrators about research, and a heightened awareness of improved client outcomes throughout an agency.*

LOCAL BARRIERS:

_____

_____

_____

_____

LOCAL OPPORTUNITIES:

_____

_____

_____

_____

THE CLIENT POPULATION

THE BARRIER: *People in drug and alcohol treatment tend to have a fairly high incidence of relapse, have high levels of co-existing disorders, and often face social problems such as unemployment or homelessness. This creates a population difficult to track and treat for extended periods of time.*

THE OPPORTUNITY: *Because our clients face such desperate situations, they are often willing to try new treatment options.*

LOCAL BARRIERS:

_____

_____

_____

_____

LOCAL OPPORTUNITIES:

_____

_____

_____

_____

Your Workbook

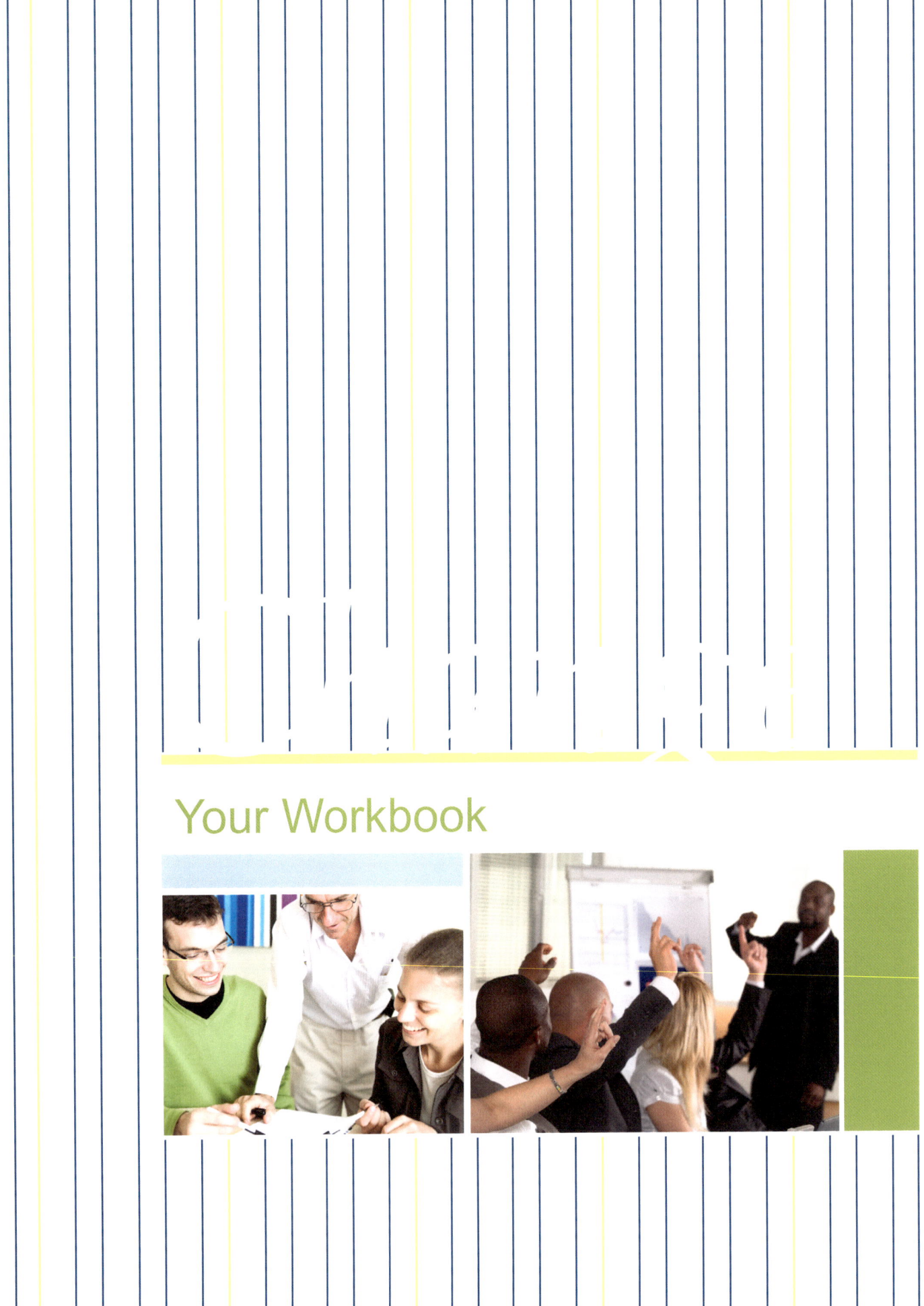

YOUR WORKBOOK

Identify the Problem

1. What is the issue or problem?

_____

_____

_____

_____

2. What data or other information support the existence of this issue or problem?

_____

_____

_____

_____

3. What is the current practice in your organization (for practitioners, administrators) that might be contributing to or maintaining this problem?

_____

_____

_____

_____

3. Use the space below to create your own questions and answers that apply to *Step* 1.

_____

_____

_____

_____

TIPS

ADD OR DELETE QUESTIONS UNDER EACH STEP BASED ON YOUR NEEDS.

ANSWER ALL QUESTIONS UNDER EACH STEP AS COMPLETELY AS POSSIBLE.

WORK STEPS 1-7 PRIOR TO IMPLEMENTING YOUR CHANGE INITIATIVE.

Organize a team for addressing the problem.

We encourage you to use a team approach from beginning to end with any change initiative. It is important to build your team with people from all levels of your agency. Your team's size will depend on the size of your organization and the particular change initiative you are implementing.

1. Who is affected by the problem (practitioners, administrators, clients, family members)? Do these individuals come from multiple disciplines (social workers, treatment counselors, mental health professionals)?

_____

_____

_____

2. What do each of these groups think about the problem? Is there any perceived need to change by each of these groups? What do they think about each other?

_____

_____

_____

3. Who are the opinion leaders within each of these groups?

_____

_____

_____

4. Who will your team members be?

_____

_____

_____

5. How will you invite team members to participate in the change initiative?

_____

_____

_____

6. When and where will you meet?

_____

_____

_____

7. How will team members communicate (meetings, memos, listservs)?

_____

_____

_____

8. How will your encourage and reward participation by team members (refreshments at meetings, recognition for participation)?

_____

_____

_____

9. Are there people from outside your agency who should be involved in the change initiative (referral agencies, funders)?

_____

_____

_____

Use the space below to create your own questions and answers that apply to *Step* 2.

_____

_____

_____

_____

_____

_____

Identify the desired outcome.

Be sure when defining your desired outcome to set goals and expectations at realistic and attainable levels. If your goals are too high and are not met, staff may become resistant to participating in future change projects.

1. **What does current research show to be a realistic outcome for the problem? (Conduct a literature review in journals, on the Web, with government sources, etc.)**

    _____
    _____
    _____

2. **How have colleagues in similar organizations addressed the problem? What approaches have they used? What has been most effective? What outcomes have they achieved?**

    _____
    _____
    _____

3. **What do staff members think would be a realistic outcome for the problem?**

    _____
    _____
    _____

4. **Reflecting on this information, what will be your desired outcome?**

    _____
    _____
    _____

Use the space below to create your own questions and answers that apply to *Step* 3.

_____
_____
_____
_____
_____

Assess the organization or agency.

1. **What is the existing organizational structure and size of your agency?**

_____

_____

_____

2. **What is the mission of the organization?**

_____

_____

_____

3. **What type of work setting is it (medical, substance abuse treatment, mental health, freestanding clinic)?**

_____

_____

_____

4. **What is the staff composition (administrators, supervisors, counselors)?**

_____

_____

_____

5. **What is the education and experience level of staff?**

_____

_____

_____

6. **What is the cultural makeup of the staff and/or clients?**

_____

_____

_____

7. What are some of the organizational barriers to change (funding, physical structure, organizational structure, policies)?

_____

_____

_____

8. What are the organizational supports for implementing change (strong desire for better outcomes, identified opinion leaders, available funding)?

_____

_____

_____

9. At what stage of change is the organization operating with regard to this change initiative (precontemplation, contemplation, preparation, action, maintenance)?

_____

_____

_____

10. Where will the resources come from to provide support for the change initiative (funding, community support, internal support from counselors and clients)?

_____

_____

_____

11. What will the adoption of this change mean at all levels of the organization? What are the benefits for administrators, supervisors and counselors?

_____

_____

_____

12. What things are already happening that might lay the foundation for the desired change?

_____
_____
_____

Use the space below to create your own questions and answers that apply to *Step* 4.

_____
_____
_____
_____
_____
_____

Step 5

Assess the specific audience(s) to be targeted.

1. Who will be targeted for the desired change (administrators, supervisors, counselors, clients)?

_____
_____
_____

2. Are there any incentives to change (for counselors, supervisors or the entire organization)?

_____
_____
_____

3. What are the barriers to change (for counselors, supervisors or the entire organization)?

_____
_____
_____

4. At what stage of change are each of these target audiences (administrators, supervisors, counselors, clients)?

_____

_____

_____

5. How will the practice(s) of those involved be affected by change?

_____

_____

_____

6. Can we identify the opinion leaders within each of these target groups?

_____

_____

_____

7. What additional support will the target audience(s) need to bring about change (e.g., training, policy changes, financial, additional personnel)?

_____

_____

_____

Use the space below to create your own questions and answers that apply to *Step* 5.

_____

_____

_____

_____

_____

_____

Identify the approach most likely to achieve the desired outcome.

1. What approach does research indicate to be effective in addressing the problem? (Again, conduct a literature review in journals, on the Web, with government sources, etc.)?

_____

_____

_____

2. How have colleagues in other organizations addressed similar problems? What has been most effective? What approaches have they used?

_____

_____

_____

3. What do staff members think is an appropriate approach to reach the desired outcome?

_____

_____

_____

4. Reflecting on the information obtained, what is the desired approach you have identified?

_____

_____

_____

5. What are your reasons for selecting this particular recommended approach?

_____

_____

_____

Use the space below to create your own questions and answers that apply to *Step 6.*

_____

_____

_____

_____

_____

_____

STEP 7

Design action and maintenance plans for your change initiative.

1. Based on the stages of change, what *Strategies* and *Activities* do you think will work best for each organizational level you plan to address?

_____

_____

_____

2. What is the timeline for your change initiative?

_____

_____

_____

3. What are the resources needed to implement these *Strategies* and *Activities* (e.g., funding for training, staff time, paper and printing?

_____

_____

_____

4. Who will be responsible for implementing the specific *Strategies* and *Activities?*

_____

_____

_____

5. How will the logistics be handled (e.g., memos, gathering baseline data, scheduling training)?

_____

_____

_____

6. How will you collect, analyze and report baseline data? Will you use an assessment? Do you have a computer? What resources are available for this process?

_____

_____

_____

7. How will you include those affected by the change in the change process (invite counselors into planning sessions, solicit client opinions, invite input from community partners, board members, family members?

_____

_____

_____

8. What evidence will be presented to the target audience(s) to support the desired change?

_____

_____

_____

9. How will the pros and cons of adopting the recommended approach – perceived and real – be presented (to clients, practitioners and administrators)?

_____

_____

_____

10. What *Activities* will be employed to maintain the technology transfer initiative (quarterly progress meetings, monthly reports on progress toward outcomes)?

_____

_____

_____

11. What resources are needed to implement and maintain this initiative?

_____

_____

_____

Use the space below to create your own questions and answers that apply to *Step* 7.

_____

_____

_____

_____

_____

_____

 STOP

and remember to work *Steps* 1-7 <u>BEFORE</u> you proceed to *Step* 8.

Implement action and maintenance plans for your change initiative.

Now it is time for you to put Steps 1 through 7 into action!

Evaluate the progress of your change initiative.

You'll use information collected in Step 9 to determine if changes to your action and maintenance plans need to be made (Step 10).

1. **As you begin implementing the change initiative, what is the initial feedback from your target audience(s)? What is the reaction to print materials, training, online courses, etc.?**

_____

_____

_____

2. **From the client, staff or administrative perspective, what adjustments need to be made to your plan?**

_____

_____

_____

3. **Have the objectives of your change initiative been met? What is the impact of your efforts?**

_____

_____

_____

4.   How will you share the results of your change initiative with frontline staff, supervisors, administrators and the research community?

_____

_____

_____

5.   How will you celebrate successes/results and support continuous feedback?

_____

_____

_____

Use the space below to create your own questions and answers that apply to *Step* 9.

_____

_____

_____

_____

_____

_____

Revise your action and maintenance plans based on evaluation results.

Now it is time to revise your current change plans based on the information you collected in Step 9. Once you have decided which revisions to make, you can continue the change process.

1. **How will you incorporate evaluation feedback into your plans?**

_____

_____

_____

_____

2. **How will you address resistance to the change initiative?**

_____

_____

_____

Use the space below to create your own questions and answers that apply to *Step* 10.

_____

_____

_____

_____

_____

_____